594. BOYLE. R.P.

MOLLUSCS A  KU-517-818

MAN

The Institute of Biology's
Studies in Biology no. 134

Molluscs and Man

P. R. Boyle
Ph.D.
Department of Zoology,
University of Aberdeen

Edward Arnold

© P. R. Boyle, 1981   PA – 019607

First published 1981   594
by Edward Arnold (Publishers) Limited
41 Bedford Square, London WC1 3DQ

British Library Cataloguing in Publication Data

Boyle, P. R.
  Molluscs and man. – (The Institute of Biology's
  studies in biology, ISSN 0537–9024; no. 134)
  1. Mullusks
  I. Title   II. Series
  594   QL504.2

ISBN   0–7131–2824–0

Photo Typeset by
Macmillan India Ltd., Bangalore.

Printed by Photobooks (Bristol) Ltd

General Preface to the Series

Because it is no longer possible for one textbook to cover the whole field of biology while remaining sufficiently up to date, the Institute of Biology proposed this series so that teachers and students can learn about significant developments. The enthusiastic acceptance of 'Studies in Biology' shows that the books are providing authoritative views of biological topics.

The features of the series include the attention given to methods, the selected list of books for further reading and, wherever possible, suggestions for practical work.

Readers' comments will be welcomed by the Education Officer of the Institute.

1981   Institute of Biology
                                        41 Queen's Gate
                                        London SW7 5HU

Preface

The Mollusca are one of the major animal phyla which interact with man in a variety of ways. They form significant world-wide food resources, which are capable of expansion through fisheries and aquaculture. But as pests, fouling organisms and disease vectors their activities are often harmful to our interests.

Highly characteristic basic features are common to the group which has evolved such diverse grades of organization as the chitons and the cephalopods. In these short chapters, my aim has been to collect some examples of the variety of molluscan studies and their practical applications, with which to amplify the traditional treatment of the group. Beginning with a synopsis of the molluscan plan I have tried to relate the applied aspects of each topic to the relevant biological features of the animals.

I am grateful to my colleagues Dr D. T. Gauld and Dr D. Raffaelli, Zoology Department, and Dr J. Mason of the Marine Laboratory, Aberdeen, for their help and critical advice on the draft manuscript.

Aberdeen, 1981   P. R. B.

Contents

1 The Molluscan Plan

The number of described species of molluscs is estimated to be in the order of 100 000, placing them second only to the arthropods as the phylum with the most species. Although large differences in the complexity of organization are apparent between the seven living classes, and adaptive radiation into various habitats has resulted in animals which look very different, the molluscs are organized upon a basic plan, the main components of which are clearly recognizable throughout. After a brief description of their common features I have selected aspects of the major classes for further discussion with which to illustrate the important differences between them.

1.1 The molluscan plan

Almost one hundred years have elapsed since the description of the Mollusca was refined to the point which finally separated them from other invertebrate groups such as the barnacles, brachiopods and tunicates. Subsequently, the accumulated work of many malacologists has shown how the essential features can be combined in a hypothetical basic mollusc, a *morphotype*. Whether this morphotype also represents an ancestral form from which the present classes have evolved, does not alter its main usefulness as a general model with which to understand the molluscan plan (Fig. 1–1).

The hypothetical basic mollusc is limpet-like in appearance having a single, conical *calcareous shell* surmounting a broad foot on which the animal creeps. This hard shell, although secondarily lost in many molluscan groups, protects the soft body and is very characteristic of molluscs. It is secreted by an important epithelial layer, the *mantle*, which covers the dorsal surface of the body. The shell grows in size only at the edges, where the mantle lays down calcium carbonate derived from the environment, into organized crystal layers within a protein-aceous matrix. Some thickening of the shell is also possible by addition to the inner, nacreous layer. The mantle edge is an important sensory and muscular region and figures strongly in the evolution of special structures such as the siphons of bivalves.

Posteriorly, the mantle encloses a *mantle cavity* which is open to the environment. The gut, excretory and reproductive systems, typically open into the mantle cavity which also contains the gills and certain sense organs. The mantle cavity thus forms a crucial area of exchange between the animal and environment. It is one of the most plastic and adaptable of molluscan features; its detailed morphology and organiz-ation is important to evolutionary studies as well as being informative about the physiology of the animal and its interrelationships with the

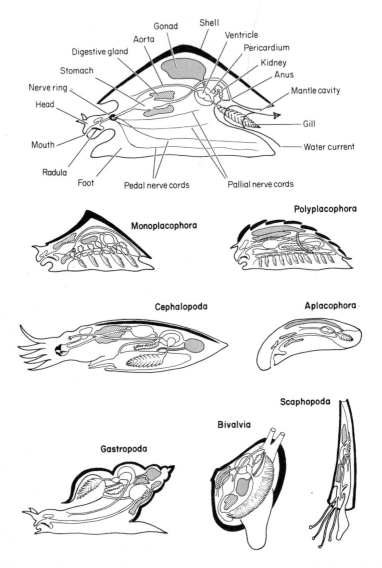

Fig. 1–1 The molluscan plan and radiation of the classes. The 'basic mollusc' illustrated represents a *morphotype* embodying the basic features of the phylum and from which the living classes may be derived. (Based on Wells, M. J., 1968, *Lower Animals*. Weidenfeld and Nicolson, London.)

physical environment. In advanced gastropods, the mantle cavity may disappear completely (Appendix).

The mouth opens into a complex buccal cavity which encloses a ribbon of chitinous teeth, the *radula* (Fig. 1–2). Only in the modern bivalves is the radula completely lacking. The basic feeding movement is a forwards and upwards movement of the radula which is then withdrawn, bringing small particles of food into the gut. Considerable detail is known about the different digestive processes in living molluscs. Probably the basic pattern is one in which cells of the diverticulae in an extensive *digestive gland* are mainly responsible for digestion and absorption, by a combination of intra- and extracellular means.

Most of the cells of the nervous system are grouped into a series of paired ganglionic masses. These are concentrated at the head end and grouped around the oesophagus into a *circum-oesophageal nerve ring*. A system of long nerves connect with peripheral regions.

Although basically *coelomate*, that is, having a body cavity, the coelomic spaces are restricted to the pericardium, kidney and gonad. Blood circulates slowly in a system of open sinuses and generally carries in solution a copper-based respiratory pigment *haemocyanin*. Haemo-globin is found in active muscles such as the radular retractors and also circulates in the blood of planorbid snails. The circulation is driven by a contractile heart, but movements of the body are probably just as important in moving the blood through the tissues. The blood has a most important function as a hydraulic fluid, for in these animals which have no hard, jointed skeleton, nearly every movement relies on the hydrostatic skeleton provided by the blood spaces and complex musculature. A urine is formed apparently by ultrafiltration through the ventricle wall into the pericardial cavity although exceptions to this may be found among pulmonates and cephalopods. The urine is drained into the mantle cavity through a nephridial canal or kidney, which is often quite elaborate and modifies the composition.

In the majority of molluscs the sexes are separate and fertilization occurs externally. A significant number are hermaphrodite, changing sex during life (see § 4.2.2, 4.4.2) and many of them have evolved sophisticated means of internal fertilization (e.g. pulmonates and cephalopods) and brooding. In marine molluscs larval development is often prolonged and planktonic.

The seven classes of molluscs with living representatives differ widely in their appearance and habits but most of the basic molluscan features can be clearly identified in each (Fig. 1–1; Appendix). Although shells are easily preserved in fossil beds, the fossil history of the group does not allow definite conclusions to be made about the ancestral forms or about evolutionary links between the present classes. From their first appear-ance in the fossil record at the end of the Cambrian, the classes were distinct and intermediate forms have not been discovered. Within some of the classes, however, the fossil history has much to offer about evolution and adaptive radiation. Evidence mostly from comparative anatomy, suggests that the phylum probably evolved from soft-bodied, unsegmented ancestors linking it to the evolution of platyhelminth and nemertine worms. For a recent discussion of molluscan evolution and contrasting views on molluscan higher systematics see SALVINI–PLAWEN (1980) and YOCHELSON (1978).

1.2   Gastropod diversity

There are probably more living species of gastropods than the total in all of the other classes. Widely distributed in all the major marine habitats, they have successfully invaded freshwaters and are the only molluscan group to establish themselves convincingly on land. Most gastropods crawl on a muscular foot but there are those which are sessile or which burrow, float or swim. Although primitively enclosed by a coiled shell, this has been lost many times during their evolution and the body form of modern gastropods is so variable as to make them the most anatomically diverse of the molluscan classes. The key to this speciation is their adaptive radiation into different feeding methods. There are filter-feeders and detritus feeders; herbivores which graze or browse; flesh-eating scavengers; those which are carnivorous on sessile animals and those which are selective predators on active prey.

The basic feeding apparatus is the radula and its complex operating musculature. It may be applied to the food in a wide variety of ways and the detailed morphology and arrangement of the radular teeth reflect their precise function (Fig. 1–2). Marine snails such as *Gibbula* and *Nerita* graze on fine algal growths with a 'licking' motion of the radula. True limpets of the genus *Patella* rasp the rock surface with the radula so firmly that rock fragments are also broken off and ingested. A small pair of jaws associated with the radula allows many gastropods to browse or nibble algal fronds or the edges of aquatic plants (e.g. the sea hare *Aplysia*). Highly specialized herbivorous feeding methods are found in those animals which selectively slit open algal cells and suck out the cell contents. *Elysia viridis* is a small opisthobranch feeding exclusively in this way on cells of the marine alga *Codium*.

A good living can be made feeding on the organic deposits present in most sheltered aquatic habitats. The mud snail *Hydrobia ulvae*, found in vast numbers on muddy estuarine shores, ingests this detritus but digests principally the micro-organisms which are richly growing there.

Other areas of the gastropod body can be used for the collection of particulate food from the plankton. A classic example is that of the slipper limpet *Crepidula fornicata* (see § 4.2.2). The beating of the gill cilia in *Crepidula* generates a current of sea-water through the large mantle cavity. A mucus net across the entrance collects large incoming particles; smaller particles drop on to the floor of the mantle cavity and the finest are sieved out from the water by the gill filaments. The food so collected is packaged in mucus and picked up from the side of the head by the protruded radula. Pelagic gastropods like *Limacina* 'swim' in

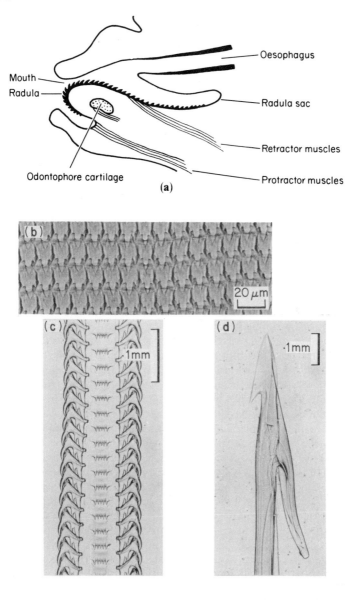

Fig. 1–2 (a) The gastropod radula in longitudinal section (based on Runham, N.W., 1963, *Q. J. Microsc. Sci.*, **104**, 271–7). The photographs show extremes of radula adaptation in (**b**) *Arion* a herbivorous terrestrial slug which has many rows of even rasping teeth, (**c**) *Buccinum* a predatory or scavenging neo-gastropod and (**d**) *Conus* where the functional radula is reduced to a single harpoon-like tooth (photographs by courtesy Zoology Department, University of Aberdeen).

surface water with the aid of flat extensions of the foot. *Limacina* collects planktonic food in the mantle cavity but related genera use the swimming folds to collect plankton, directing it to the mouth by orientated ciliary tracts.

Many groups of gastropods have evolved scavenging and carnivorous habits. Advanced prosobranchs of the order Neogastropoda, like the whelks *Buccinum* and *Neptunea*, are quickly attracted to fresh carrion such as dead fish. Other neogastropods are more active as predators. *Ocenebra* and *Urosalpinx* feed mainly on live oysters, boring a hole through the shell with the radula assisted by a softening chemical secretion (see § 4.2.1). The tropical cone shells are highly specialized carnivores in which the radula is reduced to a few harpoon-like teeth which can be rapidly stabbed at fish, worms or molluscan prey, injecting a highly toxic venom. Some species, such as *Conus textile* and *C. geographicus* are known to be capable of causing human deaths in this way.

The advanced opisthobranchs, the Nudibranchia or sea-slugs, are generally highly selective carnivores, each species feeding preferentially on a particular species of sessile animal such as a sponge, hydroid, anemone or bryozoan. Terrestrial pulmonates also show a range of food selection with slugs feeding on bulbs and roots (*Limax*), fungi (*Arion*, see § 4.4.1), or becoming specialized predatory hunters (*Gonaxis*).

Our understanding of the evolution of the primitively bilaterally symmetrical molluscs into gastropods has been complicated by two distinct processes which promote asymmetry: *spiral coiling* and *torsion*. Both of these processes occur during ontogeny, within the growth period of an individual snail, but torsion is especially important to ideas about the evolution of the class.

Spiral coiling is relatively simply understood as the packing of the mass of the body organs into a spiral coil. The coil may be a flat plano-spiral as in *Planorbis*, but more often, the spire is drawn out to the right-hand side (dextral) by varying degrees giving the typical 'snail' shape (*Helix*). Sinistral (left-handed) coiling is a relatively rare exception, but for some families of freshwater snails (e.g. Physidae, Planorbidae) where it is the rule!

The theory of torsion was put forward by malacologists to explain the anterior position of the mantle cavity in gastropods by linking it to the twisting and crossing over of certain internal organ systems. It is envisaged that the shell, its contained organs and the mantle cavity have made an anti-clockwise turn about the foot-head axis. The post-torsional result is that the paired gills are now sited above the head, together with the anus, renal and genital openings. The adaptive significance of this drastic re-organization is thought to be in allowing the larva to withdraw its head into the mantle cavity for protection and also, in placing anteriorly the respiratory and sensory structures of the mantle cavity.

1.3 Bivalve filter-feeding

Early in their history the bivalves began to use the gills for a dual respiratory and feeding role. Most bivalve gills are far larger in area than is necessary for respiration alone, their main work now being to generate a flow of water through the mantle cavity and to extract food particles from it.

Because of their considerable economic importance the filtering and sorting mechanisms of the bivalve gill have been studied in greater depth than those of other particulate feeders. There are two gill axes, one on either side of the dorsal surface of the mantle cavity, flanking the foot and visceral mass. From each axis, two rows of gill filaments hang down into the mantle cavity, each being reflected back along its own length (Fig. 1–3). In cross-section each filament pair can be thought of as W-shaped. The descending and ascending arms of each filament may be linked together by ciliary or tissue bridges; filaments may be similarly linked to their adjacent neighbours. The filaments of each gill lamella thus form a cohesive organ of minute parallel bars between which water is drawn by the beating of cilia.

Organized tracts of cilia of several types are present on the frontal and lateral surfaces of each filament. The lateral cilia are principally concerned with generating the water current through the gill. Those on the frontal surfaces transport the collected particles to food tracts on the dorsal and ventral gill margins. These food grooves terminate at the mouth where a continuous rope of food and mucus is drawn into the gut.

Tracts of especially large latero-frontal cilia sweep the space between adjacent filaments collecting food from the water current passing through. In *Mytilus edulis* at least, these are more properly called *cirri*. The cirri are each comprised of many cilia fused together along the basal part of their length but splaying apart distally, reminiscent of the way in which the barbules of some bird feathers separate towards the tip of the feather. These compound latero-frontal cilia are the central part of a

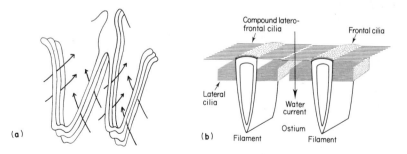

Fig. 1–3 Filter-feeding in *Mytilus edulis*. (**a**) Water flow between adjacent filaments and (**b**) the arrangement of cilia on each filament (based on an illustration in BAYNE, 1976).

very efficient filtering mechanism. Control over the filtration process is provided by adjusting the amount of water flowing through the gill and also the area of the slit between adjacent filaments which is swept by the latero-frontal cirri.

There is considerable variation in estimates of the amount of time which bivalves spend feeding, but at optimal temperatures in the natural environment, animals such as *Mytilus* probably spend as much as 95 % of the time in feeding activity. The ventilation rate, that is the flow of water through the animal, is about $1.5 l h^{-1}$ for an average 5 cm mussel. Suspended particles as small as $1 \mu m$ in diameter can be retained by the gills, and in suspensions of particles larger than $2 \mu m$, undisturbed mussels can achieve nearly 100 % retention. Experimenters using known concentrations of algal cells in suspension have attempted to estimate the proportion of filtered material actually ingested, because at high particle concentrations an increasing amount of the material filtered is rejected as 'pseudofaeces' before ingestion. Again, the estimates vary rather widely depending on the experimental conditions, but different authors report that concentrations of algal cells of between 25×10^6 cells l^{-1} and 2×10^8 cells l^{-1} may be completely cleared before pseudofaeces are produced.

These estimates for ventilation rate, particle retention and ingestion efficiency demonstrate the effectiveness of the filter-feeding method of food collection, especially in the marine environment where food concentration in the water is usually high (see § 3.1). Although energy is used to generate the flow of sea-water most bivalves require little energy expenditure for locomotion. Consequently, the bivalves which have become independent of substratum as a source of food are able to adopt a rather heavy and enclosing protective shell as well as entering the substratum itself for additional protection.

Adaptive radiation among modern bivalves has been largely a process of invading new areas which offered suitable protection for a sessile animal in which to carry on filter-feeding. Bivalves are predominantly infaunal, a range of genera such as *Tellina*, *Cardium*, *Spisula*, *Mya* and *Ensis* live within soft sediments. More specialized infaunal groups bore into soft rock (*Pholas*, *Xirphaea*, see § 4.4.1) or wood (*Teredo*, *Bankia*). The latter examples are notable exceptions to the filter-feeding habit in that these wood-borers possess cellulose-digesting enzymes and can utilize the wood as a source of food as well as protection (see § 4.4.2). Typically all of these infaunal animals maintain their contact with the water by the formation of a pair of tubes from the edges of the mantle — the inhalent and exhalent *siphons*.

Conspicuous epifaunal bivalves are the more important commercial stocks. They may be attached to the surface by the tough proteinaceous threads of the *byssus* (*Mytilus*, *Modiolus*) or fixed with calcareous cement like the oysters (*Ostrea*, *Crassostrea*). Free-living scallops (*Pecten*, *Chlamys*, *Aequipecten*) lying on the surface of sandy bottoms rely on escape swimming movements for protection. Rhythmic contractions of the adductor muscle rapidly close the shells to expel a strong water jet. With alternating opening movements due to the elastic hinge, the swimming scallop gives the appearance of a pair of false dentures 'biting' through the water. The physical requirements of these various habitats have involved, of course, considerable re-orientation of the bivalve body plan and changes in the amount of development of various structures.

1.4 Cephalopod activity

The striking feature of any modern cephalopod is the impression it gives of activity and strength. Unlike their enormous heavy-shelled ancestors, the ammonoids and nautiloids dominant in the late Palaeozoic and early Mesozoic seas, the modern squids, cuttlefish and octopus are largely shell-less (Fig. 1–4). Their muscular bodies with eight, suckered arms, large eyes, rapid colour and textural changes, conspicuous breathing movements and swimming by jet propulsion are

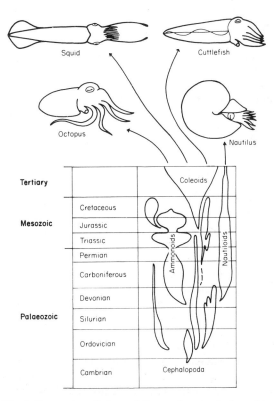

Fig. 1–4 Evolution and radiation of the modern cephalopod types. (Lower part of the diagram from PACKARD, 1972, from various sources.)

always remarkable. In many features, the evolution of the cephalopods shows operational convergence with the vertebrates and a unique degree of complexity and organization among the invertebrates.

The majority of cephalopod species are pelagic, living in the surface and middle waters of open ocean. Although early evolution of the cephalopods was in warm and shallow seas, competition with fishes probably drove them into deeper, offshore and less productive environments, the coastal waters being secondarily re-invaded by octopuses and cuttlefish.

Pelagic, surface-living squids are powerful swimmers moving in large shoals. The primitive ventilatory movements are adapted to loco-motion, the rapid expulsion of water from the muscular mantle cavity through a ventral funnel giving a form of discontinuous jet propulsion (see § 5.1). In contrast, many mid-water squids swim weakly, if at all, hanging neutrally buoyant in the water by virtue of a large fluid volume and the substitution in the coelomic fluid of the 'light' ammonium ion instead of sodium. Inshore cephalopods, mostly octopuses and cuttle-fish are largely benthic, the octopuses in particular using their arms to scramble over the bottom in addition to jet propulsion.

All cephalopods are active predators feeding on a wide range of invertebrate prey and fish. The pelagic squids catch fish but the octopuses mostly take Crustacea and other molluscs. The suckered arms are capable of a powerful grip and when prey is attacked, it is enveloped in the arms and interbrachial web and immobilized. Squid and cuttlefish have an additional pair of tentacles which is rapidly extended in a predatory strike on the prey. In addition, the prey is commonly poisoned with a secretion from modified posterior salivary glands. The ensuing dismemberment and eating of the prey is rather lengthy because only relatively small pieces can be ingested, as following the usual molluscan pattern, the oesophagus passes through the central nervous system. The radula, horny beaks, arms and suckers are probably all involved in cleaning flesh from the prey.

Many cephalopods grow rapidly, breed once and have a short life span. Daily growth increments of more than 3% of body weight have been recorded in laboratory conditions coupled with apparently high food conversion rates.

The sexes are separate and fertilization is achieved in octopuses by individual matings. The sperm is packaged by the male into a complex spermatophore which is inserted into the female mantle cavity by one of the arms (3rd right) modified for this purpose when the animal is in breeding condition. Fertilization is thus internal and the female lays her strings of fertilized eggs by attaching them to a suitable substratum. In *Octopus* species the female shows a high degree of brood care. She remains with the eggs in the hollow or home she has selected, defending them against predators, aerating and cleaning them. During this time she virtually ceases to feed and after the eggs have hatched she soon dies. Squids do not care for their eggs but often die in immense numbers after spawning which may occur in massed shoals and with general release of sperm into the water. It seems a safe generalization that each female spawns only once and that the life span of most cephalopods is only between one and two years.

Of the many special and unique features of cephalopod organization, that which most sets them apart from other invertebrates, including other molluscs, is the complexity and flexibility of their behaviour patterns. In octopuses and cuttlefish there is a wide repertoire of motor actions involving locomotion, posture, colour and texture. We can identify special behaviour patterns adapted to a variety of situations such as defence against predators, attacks on prey, mating and brooding.

To a considerable degree it seems that octopus behaviour is modified by experience. As we shall see (§ 5.3), this trait has been utilized by researchers to study the mechanisms of memory and discrimination, but, naturally it forms an important part of the survival strategy of octopus in the sea. A clear example of this is the rapidity with which a naive octopus will learn not to attack a hermit crab protected by the stinging sea anemone *Calliactis*.

1.5   The minor classes

The Monoplacophora, Polyplacophora, Aplacophora and Scaphopoda (see Fig. 1–1 and Appendix) are four classes which are minor in the sense that they are each comprised of relatively few species and have less diverse body forms than the gastropods, bivalves and cephalopods.

The class Monoplacophora was erected after the finding of a living representative of the order Tryblidiacea in deep marine trenches off Puerto Rico in 1952. This animal, *Neopilina galatheae*, and its few subsequently discovered relatives, is important in that it is a survivor of a class previously thought to be extinct since the mid-Devonian.

It is especially interesting that many of the organ systems of the bilaterally symmetrical body of *Neopilina* are serially repeated. It was the eight pairs of foot retractor muscles present in *Neopilina* which first placed it in the Tryblidiacea, because scars remaining from similar sets of muscles can be seen on the inner surface of the fossil shells. There are six pairs of nephridia and five pairs of gills. Two pairs of gonads and auricles, a paired ventricle and a ladder-like arrangement of connectives in the nervous system, all strongly suggest a segmented arrangement of the body. Such an interpretation has, of course, profound implications for molluscan evolution by linking them to the metamerically segmented phyla, annelids and crustaceans. Present opinion seems to be that these arrangements in *Neopilina* do not represent serial segmentation but a secondary repetition of certain organ systems.

Chitons, the class Polyplacophora, are mostly littoral in distribution. They are found world-wide and are particularly well represented in the western Pacific. All chitons have a dorsal shell articulated into eight overlapping shell plates. With this shell surrounded by a flexible skirt or girdle and surmounting a large ventral foot, the chitons are primarily adapted to close attachment to hard substrata, grazing on an algal film. Here, also, there is repetition of some structures in the body, the foot retractor muscles, gills and auricles, as well as transverse connections in the nervous system. It is generally thought that these repeated features are consequences of an elongate, flattened body form rather than indications of true segmentation.

The Aplacophora are an obscure, vermiform group of molluscs living in soft sediments under relatively deep water. Without a proper shell, they were for a long time grouped with the chitons on the basis of the resemblances between spicules embedded in their cuticular external surface and those in the mantle of chitons. Although clearly molluscan (they have a radula) their real relationships are not clear and they remain a rather little known class (but see SALVINI–PLAWEN, 1980).

The Scaphopoda are enclosed within cylindrical shells tapering to one end and slightly curved – the elephant's tusk shells. They are a uniform and specialized group living in sand with the pointed end of the shell protruding. Both inward and outward water currents pass through the tubular mantle cavity via this narrow shell opening. From the head end, buried in the sand, special retractile tentacles move between the sand grains picking up food particles such as foraminiferans. The scaphopods are thus micro-carnivores.

2 Molluscan Resources

Our most significant direct use of molluscan stocks is as food, but other uses, as raw material for fish meal or bait, and as an occasional source of calcium carbonate, should not be overlooked. Although most of the following discussion centres on the marine molluscs, there are fresh-water and terrestrial resources of local importance. Of the large number of molluscan species, relatively few form important resources and these species are gastropods, bivalves or cephalopods.

2.1 Exploitation since early times

2.1.1 Food

The origins of man's use of molluscs as food are lost in prehistoric time. Certainly, practically all of the coastal dwelling sites of neolithic man show evidence of his use of locally available molluscs. Shell mounds or shell middens in western Europe, the Mediterranean, Malay Archipelago, Australasia and South America testify to the generality of this exploitation. Many of these shell deposits are of enormous size and contain thousands of tons of shells of local edible species such as oysters, mussels, cockles, winkles and whelks. Human artefacts are common within these middens, and together with evidence of fires, suggest that they are indeed the result of human exploitation and have not accumulated by natural causes.

Throughout the classical and mediaeval periods many shellfish were prized as food. The Romans were largely responsible for introducing the consumption of land snails and the techniques for oyster culture into many parts of western Europe. 'Escargotieres' or snail gardens, enclosures where the snails were kept and fed, often on aromatic herbs such as marjoram and thyme, were common. In Britain a legacy of this influence is found today in the populations of *Helix pomatia*, the Roman Snail, in certain areas of southern Britain often associated with Roman settlements. Although the Romans are often credited with its introduction to Britain, *H. pomatia* has been found in pre-Roman deposits.

2.1.2 Bait

Probably as old as the direct use of molluscs as food is their use as fishing bait. Mussels and limpets especially were used extensively by line fishermen to bait hooks. Before the general decline of line fishing as a commercial fishing method in Britain there was appreciable local activity in the collection, preparation and storage of these molluscs for bait. A century ago, it was estimated that some 400 000 mussels for human consumption were sold annually at market in Edinburgh and Leith, but throughout the Firth of Forth the annual use of mussels for bait was in the region of 30–40 million. Certain areas, such as the Firth of Clyde, operated as centres for the collection and distribution of mussels and traded them over a wide area of the country. Other species such as the queen scallop *Chlamys* and the whelk *Buccinum* have a history of use as bait.

2.1.3 Dyestuff

Tyrian Purple, the purple dyestuff highly sought after and expensive in classical times was prepared from neogastropods of the family Muricidae. Chemically the dye is 4–4'–dibromindigo or 6–6'–dibromindigo and is obtained from the hypobranchial gland principally of *Murex brandaris*, *M. trunculus* or *Thais haemastoma*. When first collected it is a greenish-yellow colour which darkens to red and finally purple on exposure to sunlight.

According to JACKSON (1917) the industry probably originated in Phoenecia or Crete where taxes levied on the catch of purple-shell were mentioned as far back as 350 B.C. Excavations of banks of purple-shells date them back to at least 1600 B.C. from the presence of early Cretan objects. Manufacture of Roman Imperial Purple at Tyre began before 300 A.D. and was a state monopoly by 383 A.D.

Because of the large number of shells and the intensive labour required, purple-dyed cloth was always expensive and restricted to materials such as temple hangings and the robes of priests and nobles. It became a badge of rank to own purple cloth and its use within the Roman Empire was controlled. The excavation of banks of purple-shells, characteristically broken, at sites in Japan, China, Malaysia and pre-Columbian Central America, suggest that purple-dyeing industries became widespread throughout the world. It was still used in the 16th – 18th centuries for dyeing linen in parts of England and other European countries.

2.1.4 Shells and pearls

The uses made by man and the value which he has placed throughout history on molluscan shells and their products, the pearls, is a fascinating topic. For purely practical purposes, shell banks and shell gravel deposits have been commonly used as an easily available source of calcium for agriculture, farmers even carting off prehistoric shell middens. As crushed shell it is still a valuable additive to agricultural feedstuffs, for example, for chickens. The same material has been used traditionally to make a fine quality lime or ground up to produce tooth powder.

Cowry shells, particularly the money cowry (*Cypraea moneta*) and ring cowry (*C. annulus*) have functioned as currency in many societies. The shells were used in their natural state throughout much of southern Asia, certain Pacific Islands and across the African continent. During the period that much of Africa was explored by Europeans, shell money was widespread. In India, cowries operated as a lesser currency to metal money and in certain kingdoms there was no other currency. Even as late as 1801, revenues in certain British districts were collected in cowries (JACKSON, 1917).

As personal decoration we are familiar with pearls up to the present day. Although the majority of those coming on to the market are now cultured or artificial, very large industries and fortunes have been based on the pearl fisheries. Many molluscs produce pearls from the nacreous material which lines the inner surface of the shell (calcium carbonate usually in the aragonite form). The pearl is produced by the mantle to enclose small foreign bodies which enter the soft tissues (these may be pieces of grit, but are probably most often the cysts of parasites). The main sources of natural pearls are the marine pearl oysters of the genus *Pinctada*. The most important fisheries to have supplied the world trade were based in the Persian Gulf, centred on Bahrein Island, which at the turn of the century employed some 35 000 fishermen. The second greatest fishery was situated in the Gulf of Manoor (between Sri Lanka and the mainland of India), but there have been marine pearl fisheries throughout the world in tropical and subtropical areas, as well as significant fisheries for freshwater pearls from members of the family Unionidae (*Margaritifera*) in cooler climates.

Evolving from the pearl industry was that for 'mother of pearl', made by cutting the shell itself into the required shape. In this form the industry in shell products supplied many of the needs now fulfilled by plastics – buttons, combs, buckles, studs – on all sorts of items of dress, products which can still be seen worn by the Pearly Kings and Queens of London. Shell ornaments, amulets and anklets are familiar objects from ethnological investigations all over the world (Fig. 2–1).

In art and architecture the shell, in the form of the scallop, is a favourite decorative form from classical times to the Renaissance (COX, 1957). It has been adopted in religious art as the Christian symbol of pilgrimage, associated first with the pilgrims who travelled to visit the tomb of the Apostle St. James the Greater who was thought to have been buried at Santiago de Compostela in northern Spain. It is probably from this association that scallop is found incorporated into the armonial bearings of many of the European noble families.

2.2 The World Fisheries

Accumulated data on the world fisheries for all sorts of marine and freshwater species are summarized and published annually by the Food and Agriculture Organization of the United Nations. From these figures the utilization of different molluscan resources in various areas of the world can be compared (Fig. 2–2). In the case of shellfish, the statistics group together the products of natural fisheries with the output from aquaculture. Given in tonnes (t, 1000 kg) they record the gross weight of each species landed in each area, and thus the weight of shell is included with that of the shellfish meat. This is important to remember when

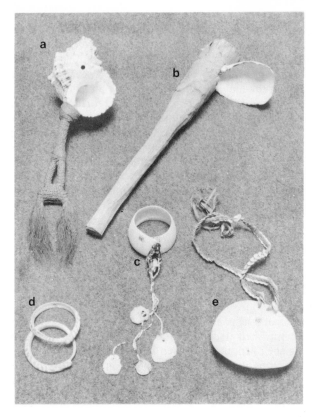

Fig. 2–1 The ceremonial, practical and decorative use of shells by human cultures in the Pacific Region. (**a**) Shell trumpet from Fiji, (**b**) digging tool, (**c, d**) armlets of *Conus* shells, (**e**) chest or pubic covering from a baler shell. (Specimens by courtesy Anthropological Museum and photographs by courtesy Zoology Department, University of Aberdeen.)

comparing the heavily shelled species, such as oysters, with the cephalopod catch. Although these figures are the best available estimate of catches on a worldwide basis, FAO caution that owing to national differences in the way the statistics are recorded, the figures are almost certainly significant under-estimates of the real catch.

2.2.1 The freshwater catch

According to the FAO (Table 1) nearly all the freshwater catch of molluscs is in Asia (Japan, Korea, Phillipines and Indonesia) with small amounts in Mexico and Fiji. It is composed of freshwater clams (*Corbicula* spp.), although small amounts of other freshwater bivalves are probably taken for food in many parts of the world. The widespread, but small scale, fisheries for freshwater pearls (mostly *Margaritifera*) are excluded as well as all of the edible terrestrial gastropods.

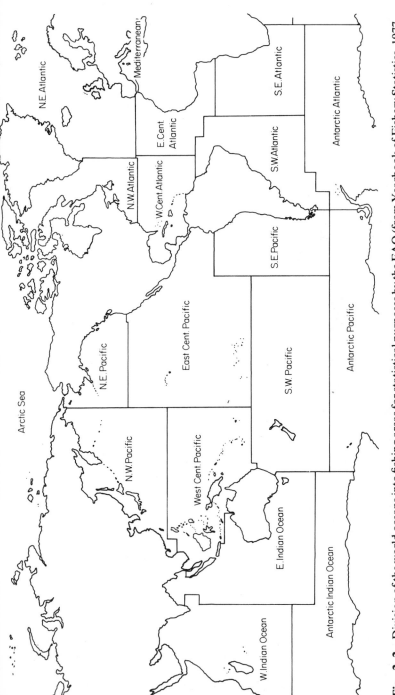

Fig. 2–2 Division of the world oceans into fishing areas for statistical purposes by the FAO (from Yearbook of Fishery Statistics, 1977, Catches and Landings, Vol. 44. FAO, Rome).

Table 1   Nominal world catch of freshwater molluscs (tonnes) in the year 1977, by area. (Data from Yearbook of Fishery Statistics, Vol. 44, Food and Agriculture Organization of the United Nations.)

Africa	< 1
America – North	1517
America – South	< 1
Asia	129 976
Europe	< 1
Oceania	1150
U.S.S.R.	< 1
Antarctica	< 1
Total	132 643

2.2.2   Marine gastropods

In 1977, marine gastropods composed less than 2 % of the total marine molluscan catch (Table 2). In the Atlantic, the north-eastern region was the most productive, with the U.K. catches of whelks (*Buccinum undatum*) and winkles (*Littorina littorea*) making up the bulk of the landings. The west central Atlantic area produced the second largest total catch of gastropods, largely of conch shells (*Busycon* spp., *Strombus* spp.) landed in the West Indies and Mexico.

Three-quarters of the total gastropods fished were from the Pacific Ocean. Half of these went to Japan and Korea in the form of abalones (*Haliotis* spp.) and topshells (*Turbo*). The bulk of the remainder (southwest, 14 199 t), mostly of *Concholepas*, went to Chile. Although not appearing in the FAO figures, as recently as 1974 it was reported that some one and a half million chank shells were fished annually in India. These were used in the manufacture of bangles and curios, in worship and for trumpet blowing ceremonials.

2.2.3   Marine bivalves

With a total catch weight of more than two and a half million tonnes, the bivalves are more conveniently discussed in terms of the major sorts. Only two genera of true oysters are involved. *Ostrea* spp. account for only about 6 % of the total. Most of these were the New Zealand dredge oyster (*Ostrea lutaria*, Pacific southwest, about 38 000 t) with the European oyster, *O. edulis* accounting for another 11 000 t (Atlantic northeast).

The majority of the catch consisted of *Crassostrea* spp. (94 %) most of it from aquaculture operations. The bulk of this total was made up from *C. angulata* in the northeast Atlantic (mostly France); *C. virginica* in the U.S.A. (northwest and west central Atlantic); and *C. gigas* in Japan and Korea (northwest Pacific). The pearl oyster fishery is not included.

Mussels (mostly *Mytilus* spp.) were derived mainly from aquaculture in Europe (northeast Atlantic), principally in Holland and Spain with substantial amounts from France and Denmark. In the Pacific it was

Table 2   Nominal world catch of marine molluscs (tonnes) in the year 1977. The fishing areas are shown in Fig. 2-2. (Data from the Yearbook of Fishery Statistics, vol. 44, Food and Agriculture Organization of the United Nations.)

Fishing area	Gastropods	Bivalves				Cephalopods			Grand totals
	Abalones, winkles, conchs	Oysters	Mussels	Scallops	Clams cockles, arkshells	Squids	Cuttlefishes	Octopuses	
Arctic Sea									
Atlantic – Northwest	878	120 571	5451	217 467	241 613	129 808			715 788
Atlantic – Northeast	9698	110 885	290 827	38 699	37 058	8129			511 598
Atlantic – West Central	2496	165 782	82	13 765	21 616	2129	10 424	5878	212 749
Atlantic – East Central		146				25 220	25 749	6879	147 718
Mediterranean and Black Sea	70	5426	11 255	214	3423	10 684	15 034	96 603	66 726
Atlantic – Southwest	37	448	4660	88	136	3554		20 620	9006
Atlantic – Southeast	736	49	12			4085	6	83	5023
Atlantic – Antarctic						1		135	1
Sub-total Atlantic Ocean and adjacent seas	13 915	403 307	312 287	270 233	303 846	183 610	51 213	130 198	1 668 609
Indian Ocean – West	3343	1	165		200	218	26 423	538	27 380
Indian Ocean – East		168		1008	966	2540	523	9	8722
Indian Ocean – Antarctic									
Sub-total Indian Ocean and adjacent seas	3343	169	165	1008	1166	2758	26 946	547	36 102
Pacific – Northwest	27 365	389 241	76 160	126 723	380 968	473 087	46 631	47 905	1 568 080
Pacific – Northeast	428	25 499	87	3325	3453	780	33 572
Pacific – West Central	2099	7998	123 960	589	119 472	46 273	81 336	4931	386 658
Pacific – East Central	7115	2933	422		5147	9474		36	25 127
Pacific – Southwest	1164	49 362	816	5684		55 515	17		112 558
Pacific – Southeast	14 199	73	25 616	1067	16 454	275		38	57 722
Pacific – Antarctic									
Sub-total Pacific Ocean and adjacent seas	52 370	475 106	226 974	134 150	525 366	588 077	127 984	53 690	2 183 717
Grand totals	69 628	878 582	539 426	405 391	830 378	774 445	206 143	184 435	3 888 428

Thailand (west central) and Korea (northwest) which accounted for most of the catch of *Mytilus* and *Modiolus*.

By far the largest catches of scallops were made on the Atlantic (northwest) coasts of Canada and the U.S.A. The bulk of the catch in the Pacific (northwest) was taken by Japan.

The final bivalve category of clams, cockles and ark shells includes a wide variety of bivalve genera, the main ones being *Arca, Arctica, Cardium, Mercenaria, Mya, Spisula, Tapes, Venerupis* and *Venus* in the Atlantic; with *Anadara, Mactra, Mercenaria, Meretrix, Paphia* and *Siliqua* in the Pacific. The U.S.A., Japan, Korea, Thailand and Indonesia were the major fishing nations in this category.

2.2.4   Cephalopods

The cephalopod catch, mostly of squid, is a significant one. The edible parts of the animal make up about 70 % by weight, an unusually high proportion when compared with fish or crustaceans. The catch of squid has increased considerably in recent years, some of the main species currently exploited being *Loligo* spp. and *Illex illecebrosus* in the Atlantic and *Todarodes pacificus* in the Pacific.

Squids are commonly identified by fishery experts as the most abundant and under-utilized group of marine food animals (GULLAND, 1971). One severe drawback to the development of the squid fishery is the difficulty in assessing the catch which can be expected in any one year because the population density can fluctuate wildly as the figures for squid landings in Scotland show (Fig. 2–3). In a good year, squid can be a very productive fishery. CLARKE (1966) estimates that 24×10^8 individuals of the common Japanese species *Todarodes pacificus*

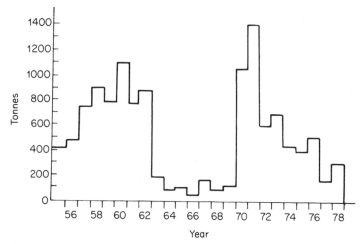

Fig. 2–3   Annual landings of squid in Scotland 1955–78 (from Howard, F. G., 1979, I.C.E.S. Report 1979/1136, Shellfish and Benthos Committee, quoted with permission).

(average weight 250 g) were caught in 1952. This is comparable with the estimated total number of herrings formerly landed by the Atlanto-Scandian fishery; in 1963, 769 000 t were landed, equivalent to 26×10^8 individuals at 300 g per herring.

2.3　Molluscs in the marine food chain

A discussion of man's direct exploitation of molluscan stocks is biologically inseparable from consideration of the place of these animals in marine food chains. In this context, it should be appreciated that molluscs are also of great importance as food for other animals which are harvested by man.

2.3.1　Gastropods

The gastropods probably evolved as epifaunal marine animals scraping plant material from rock surfaces with a broad radula of many teeth. Most modern archaeogastropods still feed by grazing in this way and some of these primary consumers, such as the limpets (*Patella*) and abalones (*Haliotis*) may be significantly exploited. The mesogastropods exhibit a wider range of feeding methods but few of them such as the winkles (*Littorina*) and conchs (*Strombus*), which graze on algae or bacterial film, are of direct importance.

Herbivorous gastropods are restricted to the photic zone where sufficient light penetrates for the growth of plant food. They are, therefore, accessible on the shore or in the immediate sub-littoral zone and are easily gathered on a local scale, without special equipment. Their collection by coastal communities is still important today although with little commercial significance.

The grazing molluscs are important elements in the shore ecosystem in the balance which they maintain between algal growth and the clear areas colonized by animals. Many of them are eaten by fish or birds but they are probably not a significant part of food chains exploited by man. The small size of these populations and the broken rocky ground on which they occur, essentially limits their exploitation to hand collecting on a local scale.

2.3.2　Bivalves

As primary herbivores the bivalves contrast strongly with the foregoing account. Colonizing rock surfaces or superficial sediment layers, their method of feeding allows the development of high population densities spread over large areas of suitable substrata. Exploitation on an industrial scale may develop and in many cases assume regional economic importance.

Our use of these stocks of bivalve primary consumers constitutes an efficient method of cropping marine production. But in doing so, we may well be competing for the food of other marine species which include considerable quantities of bivalves in their diet. Bottom feeding fish such as cod, halibut, turbot and plaice are examples of such species which are themselves human food. There is good evidence that bivalves may be particularly important to critical stages in the life history of certain food fishes, as *Tellina* is to the juvenile plaice in Scottish inshore waters.

A further consideration is the methods used to harvest bivalves from soft bottoms. Invariably some sort of dredge or trawl is dragged over the ground. To be effective, the gear must disturb the surface layer of sediment to dislodge the molluscs. The complex of interrelationships established among the epi- and infaunal residents, microbial and physical factors may be destroyed and may lead to long-term biological damage to the fishing ground.

2.3.3 Cephalopods

Like many of the commercial species of white fish, the cephalopods are carnivores. The epipelagic squids and the coastal cuttlefishes and octopuses are all active predators. Their selection of food is remarkably catholic and consists of fish, crustaceans and molluscs. The cephalopods, therefore, have a rather unusual status in the food chain because, with their short life span they will often be younger than their individual food items. This situation would be paradoxical if it was not for the wide range of prey items involved which means that cephalopod growth, certainly that of octopuses, depends on a broad base of production.

It is generally assumed that cephalopods are important items in the diet of many large marine vertebrates including commercial fish like cod and tuna. CLARKE (1977) reviews the occurrence of cephalopods in the diets of marine mammals and birds such as whales, seals, albatrosses and penguins. Quantitative estimates of the amount taken are beginning to be made from the cephalopod beaks, usually identified to the family, which are found in the stomachs of the predators. From this evidence it appears that sperm whales feed more or less exclusively on squids. Taking conservative figures for the size and composition of the sperm whale populations and the rate of food consumption measured in captive whales of other species, Clarke estimates that in 1972 at least 100 million tons of squid were eaten by sperm whales. Using estimates for the sperm whale stock of 1946, when it was virtually unexploited, these whales probably ate more than 260 million tons of cephalopods. The southern elephant seal is the most important squid-eating seal, and from middle-range estimates for weight and population, it has been calculated that they consume 10 – 20 million tons of squid annually.

These figures are all the more remarkable when they are compared with an estimated total world fish catch of 60–70 million tons per year (GULLAND, 1971). The sperm whales in the foregoing calculations are mainly eating bathypelagic or bathybenthic squid species caught during deep dives to around 1000 m. At these depths most of the squids have weak watery bodies with high concentrations of ammonia and lower calorific values than surface-living species. The two points that follow from this are, that the whales may be taking considerably greater tonnages than these estimates allow, and that they are exploiting a resource which is largely unavailable and of little interest to man directly, who exploits the muscular, surface-living squids as a fishery.

2.4 Pollution hazards

A number of molluscan fisheries have at some time been blamed for incidences of food poisoning, although often the source of the trouble could not be precisely traced (HALSTEAD, 1978). Bivalves, however, have been known for centuries to be responsible for outbreaks of illness and even death among those who consume them. The several causes of illness are now well known and they result from the accumulation in the bivalve of foreign material, gathered from the surrounding water by its efficient method of filter feeding.

2.4.1 Bacteria

The commonest hazard is the accumulation in the bivalve of bacterial pathogens which cause illness in humans. The source of these bacteria is human sewage, which is commonly present in coastal sea-water because of man's historical exploitation of the sea and the nearness of human settlements to bivalve stocks. The bacteria can live long enough in sea-water to be filtered out by the bivalve and accumulate in its gut. Serious diseases such as typhoid, cholera and dysentery can be transmitted by eating raw shellfish from polluted waters as well as milder gastro-intestinal infections. It has recently been shown that even viral diseases, such as infective hepatitis and viral gastroenteritis can be transvected in this way.

Ideally, of course, sewage pollution and shellfisheries should never come into contact, but this separation is not often achieved. In this country, local authorities have powers to prohibit or restrict the gathering of shellfish from polluted waters for commercial sale. Routine examination for the bacterium *Escherichia coli* is made to assess pollution levels. The standards require a particular source, to contain on average no more than 2 *E. coli* ml^{-1} of tissue, with only occasional samples reaching 5 *E. coli* ml^{-1}. In extreme cases, the use of the fishing ground can be completely prohibited, but the regulations usually allow gathering, provided an approved purification treatment is given. Traditionally, in the case of oysters, this involves re-laying the shellfish in intertidal pits before sale, where after flooding with clean sea-water they gradually become free of the bacteria.

A more rapid method, developed especially for mussels, involves placing the mussels in special tanks where they can be washed and placed in sterile sea-water containing chlorine (3 ppm). Sterilization of sea-water with ultra-violet light has recently come into use. The shellfish are laid in trays in a small re-circulating sea-water system and irradiated with ultra-violet light for 36 hours (AYRES, 1978). An interesting result of the use of purification for oysters has been to confer a special status on those purified, making them easier to sell. It is now common practice to submit the majority of oysters to purification whether polluted or not! Adequate cooking, usually with flowing steam, is also widely used, often preparatory to pickling and is a completely reliable way of killing the bacteria.

2.4.2 Paralytic shellfish poisoning

A dangerous form of 'natural pollution' is the accumulation in bivalves of toxins from dinoflagellates in the plankton. *Mytilus*, *Modiolus*, *Mya*, *Spisula*, *Pecten*, *Protothaca* and *Saximodus* are all edible bivalves known to be capable of transvecting this form of poisoning to humans (HALSTEAD, 1978). Planktonic dinoflagellates undergo periodic large increases of population density called 'blooms' when they may become so concentrated as to discolour the sea ('red tide'). There is a very close correlation between the concentration of these dinoflagellates (*Gonyaulax* is the main genus responsible) and the toxicity of the bivalves (Fig. 2–4).

Paralytic shellfish poisoning is specific to this dinoflagellate poison. The human response to these toxins varies greatly between individuals. The symptoms are multiple, affect the respiratory and circulatory systems and also have a powerful neuro-toxic effect which becomes evident in about 30 minutes. Death may result within 12 hours from respiratory failure (HALSTEAD, 1978). In 409 cases from the Pacific coast of North America, the mortality rate was 8.5 %. No specific antidote is known and repeated intoxications do not produce immunity.

There is no reliable method of detecting poisoned mussels (the usual bivalve involved) except for laboratory toxicity testing. The usual methods of cooking do not remove the poisoning danger although they may reduce it. The best method of prevention lies in a strict adherence to the local quarantine regulations in force.

2.4.3 Industrial pollutants

It has long been known that mussels and oysters can cause poisoning due to their accumulation of toxic metals such as copper. Mussels grown on the sides of docks and piers and copper bottoms of ships have been polluted in this way. LOVELL (1867) recommended a simple test for excessive copper in oysters . . . . . .'an ordinary needle was thrust into the green part of the oyster, and then the mollusk was immersed in pure vinegar. When copper was present, thirty seconds sufficed to cover the portion of the needle embedded in the oyster with a red coating of copper.'

A chilling example of the reality of the dangers from the accumulation of pollutants in modern industrial society is the so-called 'Minamata disease'. First reported in 1953, this disease affected people of several small coastal villages bordering Minamata Bay, Kyushu, Japan. In the original outbreak (1953–56), 52 victims were identified, of whom 17 died and most of the remainder were permanently afflicted (HALSTEAD, 1978). The illness was traced to poisoning by methyl mercury

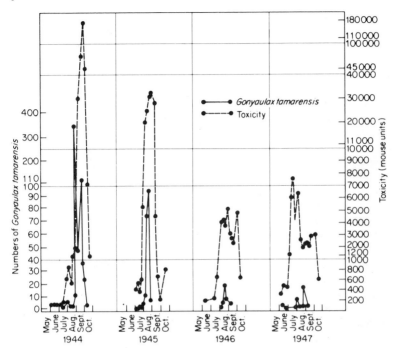

Fig. 2–4 Graph showing the close relationship between the occurrence of the dinoflagellate *Gonyaulax tamarensis* and toxicity of the mussel *Mytilus edulis* at Head Harbour, Bay of Fundy, 1944–46 (from HALSTEAD, 1978, after earlier authors).

(CH$_3$—Hg$^+$) which caused severe damage to the central nervous system, as well as congenital effects such as mental retardation, in infants born to mothers with Minamata disease.

The source of the mercury proved to be a chemical and fertilizer plant near the coast, manufacturing vinyl chloride, which discharged its waste into the Bay. Although it was not a commercial fishing area, local families took shellfish and fish for their own consumption. These animals had concentrated methyl mercury to very high levels, probably via anaerobic bacteria in the detritus and mud on the floor of the Bay. There have been several subsequent outbreaks of heavy-metal poisoning in Japan by the same route and estimates of the numbers of people who may be showing less severe, but none-the-less toxic symptoms, run very high.

Industrial pollutants of all sorts, inorganic and organic, can be concentrated in bivalves and other edible marine species. Clearly, pollution is a real, and potentially major threat to the health of man and the shellfisheries.

3 The Aquaculture Business

In spite of our long-established use of the sea for food, figures given by KORRINGA (1976) suggest that its actual contribution is a very small proportion of the potential when compared with terrestrial production. Taking into account the relative areas of land and water and the estimated figures for primary production, it can be concluded that the total production of organic matter by the photosynthetic activity of green plants must be similar on land and in the sea. Significantly, though, only about 1 % of man's food is currently derived from the ocean.

We exploit much of our marine food resources by hunting marine animals high in the food chain and far from the shore. The escalating costs and diminishing returns of this activity are stimulating more interest in the development of improved techniques for *aquaculture*. Molluscan aquaculture involves the improvement of production from natural stocks and their establishment in new areas. Most of the present methods are properly described as semi-culture because they do not involve complete management of the animals from egg to marketable product, but experiments are in progress with completely artificial rearing and feeding systems.

3.1 Molluscan attributes for aquaculture

The filter-feeding bivalves, as a group, have a number of attributes which suit them for aquaculture techniques. Compared with most fish, especially the salmonid fish popular in aquaculture, bivalves are low in the food chain. The average conversion rate at each stage of a food chain (the proportion of food consumed which is incorporated into growth), is only about 10 %. Consequently about 90 % of organic production is lost at each stage in a food chain. The energy that this represents is used in the metabolic requirements of the organisms, or enters other nutrient cycles through mortality and reproduction. The fact that nearly all of the organisms taken from the sea are fish, at the 3rd, 4th or even 5th link in the food chain is part of the explanation of why such a small proportion of our food is marine in origin. This is not an energy efficient level at which to crop the sea.

On land about 85 % of man's food is harvested as vegetable matter, but the remaining 15 % of animal products, meat, milk and eggs, are obtained from primary consumers, the second link in the food chain; even so, this animal production requires an input of more vegetable matter than man himself consumes from the first link. Of course, the bulk of the plant material used by the herbivores is useless to man directly and so this process is largely one of upgrading vegetable matter to animal protein.

In the sea, the first link in the food chain is mainly phytoplankton and the direct exploitation of this primary production by man is difficult for a number of reasons. Much of the plant material is indigestible. The size of the individual planktonic organisms is very small, which means that harvesting and processing would be difficult and costly. Probably the most significant disadvantage is that the phytoplankton is dispersed and patchy, occurring in a layer 30–100 m thick over the whole surface of the ocean. There is a small additional input of primary production from the large coastal seaweeds, but only a minute amount of this is harvested as food.

The flesh of most bivalves is palatable and rich in protein, lipid and carbohydrate. As filter-feeders and the second link in the food chain, they provide an efficient means of tapping marine organic production by converting micro-algae into usable animal food. The earlier discussion of bivalve filter-feeding (§1.3) outlined the effectiveness of their mechanisms for clearing water of particles down to bacterial, or even viral dimensions. An especial practical advantage of filter-feeders for culture techniques is that their food is replenished simply by water circulation and does not need an elaborate delivery system.

The cultured bivalve species occur naturally inshore and in shallow water, obviously convenient for access and daily working. Most of them are mussels or oysters which have two further distinct advantages. These groups are adapted to tidal conditions and the consequent exposure to air, so mussel and oyster beds can be worked partly from the shore and the harvested animals can survive in air for a considerable time. They are also normally attached to a hard surface by proteinaceous byssus threads (mussels) or a calcareous cement (oysters). By this feature their numbers can be vastly increased and their distribution controlled through the provision of suitable artificial surfaces for attachment. A great deal of molluscan aquaculture technique depends on these two factors.

Most bivalves produce enormous numbers of gametes which result in dense accumulations of planktonic larvae. Fertilization typically occurs in the sea-water but oysters of the genus *Ostrea*, retain the eggs in the mantle cavity of the female where they are fertilized by sperm drawn in by the inhalent water current. *Ostrea* also changes sex, functioning first as a male and then as a female, alternately throughout its life, spawning as both sexes even within the same year. After a short period of development fueled by the yolk of the egg, mussel and oyster larvae spend a period of growth and development in the plankton, up to 6–8 weeks depending on temperature and food supply, until the larva is ready to settle. Settlement is a crucial stage in the life of any sessile animal and a degree of substratum selection is exercised. Surface texture and adsorbed chemicals are especially important factors in promoting settlement. Inevitably these species tend to settle gregariously.

The key features of bivalve development for the aquaculturist are: that very large numbers of eggs are produced (one million per spawning female is a common estimate), the resulting larvae settle densely within a short space of time, the substratum selection operated by the settling larvae allows their collection by the farmer. Once settlement has been achieved, the larvae adhere firmly to the surface and metamorphose rapidly into young mussels and oysters. These newly-settled organisms are called *spat* and the process of settlement is the *spatfall*. After a short period of growth the tiny animals are referred to as *seed* and may now form the raw material of a culture operation. There is an increasing trend towards short-circuiting the natural cycle of fertilization, development and settlement, through the supply of seed oysters from hatcheries. Brood oysters are kept to supply ripe gametes and ensure high rates of fertilization; after fertilization the developing larvae are fed on controlled suspensions of algal cells and high rates of survival and settlement success can be achieved. As well as reducing mortality and eliminating much of the chance element of successful fertilization and settlement, theoretically, at least, genetic selection of the stock can be practised.

Bivalve aquaculture also has a number of drawbacks. The large number of eggs produced means that each is very small, about 150 μm in diameter in *Ostrea edulis*, and as there are practical difficulties of handling eggs of this size the hatchery production of seed is costly. Natural spatfall may be quite unpredictable in density and will vary from year to year for many reasons. The susceptibility of bivalve populations to pollution has already been discussed (§ 2.4) but it is worth recognizing that this problem may become more acute when molluscan culture is considered. For economic reasons such as availability of labour, the reduction of transport and processing costs and the proximity of markets, aquaculture enterprises would develop naturally in suitable shallow waters close to centres of population. It is precisely the same areas which are most likely to become polluted from domestic and industrial sources and to be under pressure from alternative uses such as harbour construction or amenity use. This conflict of interests is apparent in the difficult legal framework within which coastal aquaculture has to operate.

Lastly we must recognize that investment in aquaculture projects is a commercial operation and that a lack of sufficient markets is a brake on development in some areas. Partly, this is due to reasons of taste or to concern over pollution and disease risks, but also to the relatively high labour input which may keep the cost of cultured shellfish high. For these reasons it is likely that shellfish production from aquaculture could proceed more readily in less developed countries where pollution risks are lower and labour intensive industry is welcome.

3.2 Culture of mussels and oysters

Species of relatively few genera are involved in large-scale culture for food: three of major importance are flat oysters (*Ostrea*), cupped oysters (*Crassostrea*) and mussels (*Mytilus*). Of course, the natural populations of these bivalves have been exploited for food since prehistoric times

(§ 2.1), but there is considerable evidence that, at least in classical Rome and possibly China, man had begun to switch from fishing to farming some 2000–3000 years ago. Present culture methods, for oysters especially, are thus part of a continuous evolution of culture techniques and adaptation to local conditions.

In Europe, traditional culture methods for *Ostrea edulis* are found in most countries with suitable coastlines, such as the Netherlands, France, Britain, Italy and Yugoslavia. Probably these methods owe something to an early Roman influence. Relatively more recent is the culture of cupped oysters of the genus *Crassostrea*. The Portuguese oyster (*C. angulata*) in Europe, the Atlantic oyster (*C. virginica*) in America and the Pacific oyster (*C. gigas*) in Japan, Europe and North America now contribute a much greater production of protein worldwide than does *Ostrea* and its relatives. The commercial culture of mussels (*Mytilus* or closely-related genera) is more recent than oysters but it is well established and profitable in Europe. While Japan is still probably the world leader in molluscan aquaculture, in terms both of the variety of species which are cultured as well as total production (mostly *C. gigas*), more and more countries are experimenting with aquaculture techniques to improve their molluscan fisheries.

Great improvements can be made by the adoption of simple methods for the collection of spat or seed and the protection of growing stages from predators, parasites and adverse physical conditions. In his review of molluscan aquaculture, KORRINGA (1976) gives impressive figures for the scale of the improvement which is possible. The natural oyster beds of the Oosterschelde in the Netherlands were yielding about half a million oysters annually in 1870 when the Dutch Government withdrew permission to fish them, and instead leased plots of the extensive intertidal banks to prospective oyster farmers. About 100 years later the annual production from the same water was running at about 30 million oysters. Exploitation of the full depth of the plankton layer by suspending cultured molluscs from rafts or attaching them to poles and fences can achieve even more spectacular results. In the late 1940s the culture of mussels suspended from rafts was introduced along the deeply indented coast of Galicia, north-western Spain. The area, which previously produced no marketable mussels, 25 years later was producing around 150 000 tons of excellent mussels annually and Spain became the largest mussel producer in the world, although this position has now been lost to the Netherlands.

The traditional culture methods for oysters and mussels involve several distinct processes. These are, collection or production of the spat or seed; the growth and fattening of the product to marketable size and quality; its harvesting and processing. Details of methods vary with species and local conditions. Although it is possible to produce quantities of bivalve larvae from a number of species in hatchery conditions, the process is expensive and most current commercial operations collect the natural spatfall. One of the most important limiting factors on the production of natural beds is the restricted area suitable for larval settlement.

3.2.1   Mussels

The Dutch mussel industry relies on the collection, by dredging, of seed mussels from natural beds where they are present in enormous numbers. The seed is then re-laid on to growing beds, thus allowing the control of stocking density to improve survival and growth rate. These beds are regularly tended by hand over the 2–3 years which it takes to produce mussels of marketable size (average shell length 5.5 cm). Raking ensures the even distribution of the mussels and removes predatory starfish and crabs. The crop, up to 8 kg m^{-2}, is harvested by dredging and re-laid temporarily on to cleaning beds of firmer ground where most of the silt is discharged from the mussels. From there they can be quickly lifted by dredge and marketed live. Clearly this method requires extensive intertidal flats with suitable sediment and hydrological conditions as well as a significant tidal range to allow both access by boat and sufficient working time on the drained surfaces.

In France it has been the practice to plant rows of poles into the bottom or to stretch long parallel ropes of cocoa-fibre over horizontal supports to provide surfaces for the collection of spat and growth of seed mussels. After a few months the seed is harvested and then packed into narrow cylindrical nylon nets which are spirally wound around vertical poles (bouchots) set in the intertidal zone (Fig. 3–1). The young mussels

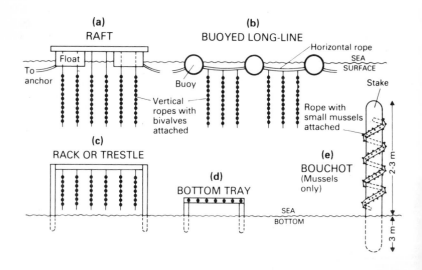

Fig. 3–1   Some of the culture techniques used in the sea for bivalves, not to scale (from REAY, 1979).

move through the net meshes and attach themselves with byssus threads. The mussels grow and fatten on these poles and begin to be harvested the following year. Arranging the mussel stock vertically in the water column has many advantages. The size of the stock which can be accommodated in a given area is increased, the area suitable for use can be greatly extended and the mussels are cleaner and more free of predators. Distributed vertically in the intertidal they can make more thorough use of the available food supply, but even so, it has been found that overcrowding on the poles can lead to malnutrition.

The advantages of using a greater depth of water and raising the mussels clear of the bottom can be most clearly seen in the suspended culture methods practised in the deep, sheltered inlets of northern Spain, the Galician rias. Mussel seed from natural settlement, mostly on intertidal rocks, is collected by hand. It is attached to ropes some 6 m to 9 m long using a specially made 'bandage' of netting, spirally wound around the rope and seed. These ropes are then suspended in the water from large, moored rafts (Fig. 3–1). Clean water, warm temperatures and rich phytoplankton contribute to very high growth rates, allowing them to grow to as much as 8–10 cm length in about a year. This high growth rate necessitates some thinning out, which is done by lifting the ropes by hand or using a hoist on the raft or a boat. Each raft carries between 600 and 1000 ropes and will produce between 30 000 and 90 000 kg of mussels annually. Compared with other species in aquaculture and various terrestrial enterprises, mussels are capable of exceptionally high productivity (Table 3). A further consequence of being off the bottom is

Table 3 Production of mussel meat (calculated as wet weight of pure meat assuming 30% meat per mussel) from various forms of mussel culture and at different locations. (From Kinne, O. and Rosenthal, H., 1977, In *Marine Ecology*, Ed. Kinne, O., Vol III, part B.)

Species	Country	Annual yield (tons/ha)	Culture method
Mytilus edulis	France	2.25	Pole culture
		12.30	Pole culture
		28.00	String culture
		42.00	String culture
	Spain	158.00–300.00	Raft culture, maximum Atlantic coast
		21.00	Raft culture, Mediterranean Coast
	Netherlands	7.7–21.0	Bottom culture
Mytilus galloprovincialis	Italy	4.0	Bottom culture
Mytilus smaragdinus	Phillipines	125.0	Bottom culture
Mussels (unspecified)	Thailand	54.0	Bottom culture

that the mussels are little troubled with predators and parasites and the few competing filter-feeders such as ascidians can be removed during the thinning operations.

3.2.2   Oysters

The farming methods in use for European oysters are generally more complicated and labour intensive than those for mussels. This reflects the present high value placed on these shellfish as luxury food items. Up to about 100 years ago, natural beds in Britain and on the continent of Europe yielded vast numbers of oysters and they were an important and cheap source of food. At Billingsgate market alone, over 495 million oysters were sold for the year 1864 (YONGE, 1966). A combination of overfishing and disastrous attacks by predators and parasites (mostly introduced with *Crassostrea* from foreign beds in an attempt to replenish the stocks) almost destroyed these natural fisheries. Only the persistent popularity of the oyster and its increasing price due to scarcity has led to a thriving industry in oyster culture, mostly in the Netherlands, France, North America, Australia and Japan. To a large extent the culture of *Crassostrea* has replaced *Ostrea*.

Oyster farming methods lay great stress on the collection of spat. Much of scientific research has gone into the life cycle of the oyster and the factors which ensure successful spatfall. It is in the careful application of this knowledge to his local conditions that the skill of the oyster farmer is most apparent. Although there are many variants of oyster farming techniques in different parts of the world, a brief description of the traditional methods which have been used for *O. edulis* and *C. angulata* in western France will illustrate the main principles.

Collection of spat is undertaken usually by placing carefully compiled stacks of roofing tiles, lime-coated to promote attachment, in selected areas at particular times. Conditions for successful spatfall in oysters are rather circumscribed and so the exact location and timing of placing out the 'collectors' is most important. On natural beds, old oyster shell is a very attractive settlement substratum and the oyster farmer may use strings of old shells suspended in the water as collectors. Successfully settled collectors may be stored in special ponds or enclosures until the following year. The young oysters are now removed from the collectors and spread out on to cleared grounds which are often fenced with wire or plastic mesh. The young oysters are carefully tended by hand to ensure their even distribution and to remove predators. At about 18 months old they are lifted and re-laid again on deeper grounds where they will lie undisturbed for at least 2–3 years until they reach marketable size and quality. The highest quality oysters are usually produced after a final transfer to specially constructed fattening ponds.

Natural oyster beds are limited to areas of successful spatfall which are not necessarily the ideal grounds for high growth. The multi-stage process outlined above allows the oyster farmer to use the total areas of intertidal and sublittoral ground available to him to maximum advantage. The collection of large amounts of densely-settled spat and its distribution over much larger areas, suitable for growth and final fattening of the oyster, obviously requires skill and experience for optimum production. Specialization allows a particular farmer to concentrate on one stage of the process, perhaps in spat collection or the production of 18 month oysters, analogous to specialization in the modern farming of cattle and sheep. In spite of the care of the oyster farmer, aquaculture, like any agricultural venture, is susceptible to a variety of hazards, both physical and biological. Storms, violent salinity changes, severe winter cold or summer drying and variations in silt deposition can all be devastating. Pollution, both chemical and microbiological, is an increasing problem and the inroads of predators, parasites, competitors and disease are, of course, especially important to the extensive culture of a single species as practised by the molluscan farmer.

Apart from Japan, where a wide variety of marine molluscs are cultured, there is little human intervention that could be called farming in the production of molluscs other than mussels and oysters. An exception might be the increasing culture of scallops and the re-stocking of natural beds which is well established in Japan and is now being tried in Europe.

3.3 Cultured pearls

The molluscan shell is formed by the activities of the mantle, concentrating and precipitating crystalline calcium carbonate derived from the sea-water. Although shell grows naturally only at the edges, many molluscs have a limited ability to repair damage in old shell. A further expression of shell-forming ability is the production of natural pearls. Foreign bodies present in the tissues are enveloped by an inpushing of mantle tissue and layers of nacreous shell are deposited around the object.

Pearls occur naturally in a wide variety of bivalves and are not always appreciated. The harvest of mussels, for example, from the shallow parts of the western and northern Black Sea is seriously harmed by the widespread occurrence of small pearls in them, making them unfit for sale for human consumption. In this case the nuisance is probably due to a trematode parasite, pearls forming around the parasitic cysts. Large natural pearls have been sought after for centuries and were collected by diving throughout the tropical Indo-Pacific and western Pacific (§ 2.2). The discovery by the Japanese that fine pearls could be induced by the artificial introduction of foreign material into the mantle of certain bivalves led to the rise of a totally new cultured pearl industry based largely in Japan.

The recipient bivalves are the so-called pearl oysters, *Pinctada* and *Pteria*, which naturally have a thick and glossy inner nacreous layer to the shell. The oysters are carefully cultured in trays suspended from rafts in sea-water. The formation of the pearls takes between 3 and 5 years, some 500 million being produced annually. The high price of the product justifies elaborate culture methods. There is considerable investment in materials and in management techniques, the rafts and their suspended trays even being towed on annual migrations along the coast in order to take advantage of the best plankton.

3.4 Conclusion and potential

Theoretically, the exploitation of marine filter-feeders low in the food chain makes sense in terms of increased production from the sea. Practically, bivalve molluscs are easily handled and adaptable to simple culture techniques. Cultured oysters and mussels show spectacular increase in production over natural beds. Although the process of transfer of man's activities from fishing to farming is well advanced, the present commercial enterprises should really be described as semi-culture with four main operations:

(a) the enhanced production of spat by providing increased areas of artificial surface for settlement, and controlled spat production by hatcheries;

(b) the transfer of young bivalves to new grounds, both natural and artificial, more suitable for growth and fattening owing to the control of stocking density and the use of food supplies not available to natural beds;

(c) the protection of the crop from physical and biological hazards;

(d) the re-stocking of natural fishing grounds (scallops).

On a small scale it is possible to culture many bivalves throughout their life cycle in hatchery conditions. Much is known about the factors which induce spawning, the conditions for development of the eggs and the growth of larvae on controlled diets of planktonic algae. These experiments cover a range of oysters, mussels, clams and scallops and it is possible to buy the seed of a number of species from hatcheries. Similar work is available for herbivorous gastropods such as *Haliotis* as well as some species of octopus. The aquaculture potential of certain octopuses (e.g. *Octopus maya*) is not so remote as it first appears – they have large eggs, very high growth and food conversion rates, can be fed on 'junk' crustacea and fish, adapt well to high stocking densities and have few disease problems.

Obviously great potential still exists for the adaptation of the present simple aquaculture techniques to new areas of the world. In many countries, suitably sheltered areas of water exist where these methods could be usefully applied. One block to the development of molluscan aquaculture is 'market resistance'. For religious reasons or concern over pollution and disease risks, the consumption of shellfish may be limited. Significant steps need to be taken in the marketing and presentation of shellfish food, but more than anything, the assurance is needed that the proper standards of microbiological purity are being applied. Countries in which shellfish are an important part of the diet, increasingly expect suppliers to give such guarantees.

In the long term we may expect that hatchery techniques will allow greater selection of improved strains for growth and disease resistance. While the use of surplus heat from power stations could be usefully applied to raise temperatures and growth rates, other forms of inshore pollution must be increasingly controlled if molluscan aquaculture is to thrive. The technical feasibility of completely closed aquaculture systems for molluscs has been experimentally demonstrated. Nutrients, supplied from secondary sewage treatment, are mixed with sea-water and used to grow a crop of unicellular algae which is then fed to filter-feeding bivalves. Even the sea-water can be partially re-cycled. Although attractive in theory, such systems have yet to prove themselves commercially viable and able to control public health hazards such as the transfer of sewage associated viruses.

4 Pests and Disease

The preceding chapters have considered the molluscs as sources of food and various products for man and other animals. Many molluscan species, however, achieve great significance to man as pests and as vectors of serious disease.

4.1 Pests of agriculture

4.1.1 Terrestrial slugs and snails (Pulmonata)

Although there are land-living forms of several molluscan groups (e.g. the large and successful mesogastropod family, Cyclophoridae), only the pulmonate gastropods have been really successful in terrestrial habitats. There are many thousands of species of these slugs and snails, far more than the freshwater colonists. Many of these species have distributions restricted to islands or outcrops of ground on which they have become isolated. Many others have wide natural ranges and will readily spread into new territory as a result of artificial introduction (§ 4.1.2).

The rich and varied plant life of the land has obviously provided an opportunity for considerable speciation and radiation among these predominantly herbivorous molluscs. The Pulmonata are probably polyphyletic in origin, the common features of adaptation to terrestrial conditions having evolved several times. The basic gastropod arrangement of gills mounted in a mantle cavity, through which oxygenated water was circulated, has been lost. The gills have been functionally replaced by modification of the mantle cavity itself to form an air-breathing lung with a densely vascularized roof to the cavity. In typical snails such as *Helix* and most slugs, the greater part of the mantle margins are fused, leaving a restricted aperture, the *pneumostome*, on the right hand side. The muscular floor of the cavity causes regular breathing movements, coupled, in *Helix*, with rhythmic opening and closing of the pneumostome. These features presumably reduce the rate of water loss by evaporation from the lung.

Although covered in mucus and losing moisture by evaporation, it appears that the skin of slugs and snails does not allow water to be transferred as rapidly as predicted and the presence of some sort of permeability barrier is postulated. In spite of these adaptations and although some species can live in the hottest and driest of desert environments (e.g. *Sphincterocheila*), slugs and snails are generally dependent on rather humid conditions and usually become active only at night.

Slugs attack a wide range of agricultural and horticultural crops throughout the world. In temperate regions the most affected crops are wheat, potatoes and the brassicas and in Britain the most important of these pests are *Deroceras reticulatum*, *Arion hortensis* and *Milax budapestensis* (HUNTER, 1978). Whether their density reaches that of pest status depends on the favourable (for the slug) combination of circumstances such as climate, shelter and food.

Figures due to Strickland (reported in RUNHAM and HUNTER, 1970) illustrate the scale of crop loss which is possible. Slugs were the third most important potato pest in England and Wales, causing an estimated loss of 3300 acre equivalents, or 36 000 tons annually, compared with an average annual consumption of 400 000 tons. For winter wheat it was estimated that 41 000 acre equivalents were lost (not quite 2% of the total but more than that lost to other well-known pest groups such as wireworms, wheat bulb fly, rabbits and hares or cereal root eelworm). Horticultural damage possibly equalled or exceeded that to agriculture. Slugs also transmit plant diseases, fungal spores passing through the gut and transferring from plant to plant.

Control of slug populations is not easy due to the difficulty of applying effective molluscicides to them and the uncertainty of predicting the timing and place of serious infestation. The most widely used molluscicide is metaldehyde, often mixed with bran as a bait. But on an agriculture scale, the more effective control is by crop and soil management. Reduction of the surface shelter available to the slugs, particularly by removing the surface trash crop after harvesting, restricts the ability of the slugs to survive a dry spell of weather and breed successfully. Compacting the soil, to reduce cracks and pores, also has some effect but it is rarely a desirable method on heavy soil where slugs are most common. In some cases, repeated rotovating between crops will reduce the slug population.

4.1.2   The Giant African Snail Achatina fulica

One particular land snail has achieved worldwide pest status through human introductions. It is of major economic importance and the subject of intensive research (MEAD, 1961). *Achatina fulica* is a very large achatinid snail, commonly reaching 15 cm in overall length for a shell length of about 10 cm with the maximum recorded being about twice these figures (Fig. 4–1).

The snail originated in a relatively restricted area of East Africa where its generally omnivorous habits and modest population density apparently caused little problem to local agriculture. As the map shows in Fig. 4–2, from the late eighteenth century onwards the snail began to disperse throughout a sequence of countries across the Indian Ocean, the Malay Archipelago and onwards across the Pacific Ocean. MEAD (1961) reviews the available evidence about these introductions and concludes that the dramatic increase in range was brought about by human agencies both accidental and deliberate. He points out that the effect of World War II in the Pacific was to greatly accelerate its spread. Abandoned crops throughout the region assisted its establishment and breeding, while the movements of large amounts of war materials and salvage throughout the area transferred the snails to many otherwise inaccessible spots.

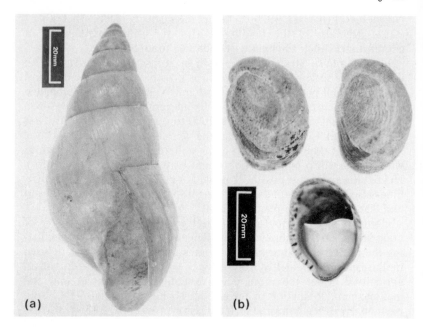

Fig. 4–1 Molluscan pests (**a**) *Achatina fulica*, (**b**) *Crepidula fornicata* (photographs by courtesy Zoology Department, University of Aberdeen).

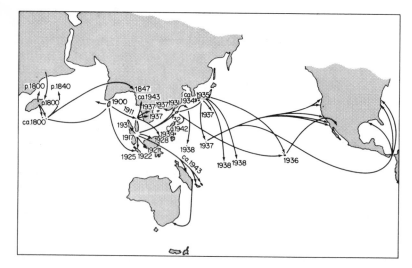

Fig. 4–2 Dispersal pattern of *Achatina fulica* from East Africa. The map shows the routes by which the snail has probably spread, although in some cases interception by agriculture authorities have prevented its establishment (from MEAD, 1961).

In many of these tropical areas, *A. fulica* rapidly bred and built up population densitites of pest proportions. It attacks a wide variety of horticultural and agricultural crops and causes severe economic damage, usually in rural areas where income from agriculture is particularly important.

Many different measures to control its numbers have been attempted. Molluscicides may be effective when they can be applied directly to the animals, but the cryptic habits of this and most snails negates their widespread application. *Achatina* has few natural predators and is seldom favoured as human food. Extensive programmes of biological control have been attempted in the Pacific and in some islands, diseases, parasites and predators have all been introduced. Probably the most widely introduced predators are gloworm beetles such as *Lamprophorus tenebrosus* the larvae of which feed voraciously on *Achatina*. Natural predators from East Africa, notably the streptaxid snails *Gonaxis* and *Edentulina* have been introduced to many new sites in an effort to reduce the density of *A. fulica*. It is probably true that most of these introductions have had no overall effect in controlling the spread or numbers of the snail. Indeed, they provide classical examples of the problems of biological control. Up to 1961 (MEAD) a total of 9 predatory beetle species, 2 presumed parasitic flies and 8 predatory snails had been introduced into Hawaii in attempts to control the nuisance. As with the smaller agricultural pests, it is the combination of good husbandry methods and reduction of ground shelter for the snails which will prove to be the best long term solution. To follow this topic up to date, and in greater depth, see MEAD (1979).

4.2   Pests of aquaculture

Like any sort of farming practice which involves the extensive cultivation of a single species, aquaculture operations are prone to a number of pests. Contributing significantly to the failure of certain European oyster farming areas are two molluscan pests, a predator and a competitor, a situation which is repeated in many parts of the world.

4.2.1   *The American oyster tingle* Urosalpinx cinerea

The neogastropods, as a whole, are active predators or scavengers. Members of the family Muricidae, in particular, specialize in feeding on bivalves which they attack by boring a hole through the shell. In most parts of the world, one or several species of this family are pests of commercial shellfish beds.

In Britain, the European or rough sting winkle (*Ocenebra erinacea*) has preyed on oyster beds to some extent, but it was the accidental introduction of the American oyster tingle (*Urosalpinx cinerea*) which led to severe effects on the oyster beds of Kent and Essex. It was first recognized in 1928, although it must have been here several years earlier, probably confused with *Ocenebra*. The oyster tingles must have arrived with oysters brought from America for re-laying. Although it will eat a variety of food species, it became very abundant on oyster beds where it was unknown 15 years earlier and where up to 10 000 per acre have since been recorded (HANCOCK, 1962).

Believed to have been introduced at two centres only, *Urosalpinx* has not spread very far because it lacks a pelagic larval phase. In the spring of each year the oyster tingles move shorewards and lay a series of egg capsules under stones from which about ten young tingles will hatch. The newly-hatched young eat oyster spat and become mature in about two years. *Urosalpinx* bores the oyster shell by a combination of abrasion by radular movements and a softening secretion from an accessory boring organ. Drilling takes a considerable period, perhaps two days for an oyster 5 cm in diameter and a full week for one 10 cm across. Once the hole is completed, the proboscis is inserted and flesh rasped from within the oyster by the radula.

Considerable damage to oyster beds in Britain and America has been caused by *Urosalpinx* and a proportion of the past decline in oyster landings here and in America can be attributed to its activities. In Essex rivers it is estimated that tingles destroy half of the annual oyster spatfall. To prevent their spread, government controls prohibit any unlicensed transfer of shellfish stock to new grounds and the introduction of shellfish from abroad. Similar molluscan pests affect most oyster cultivation. Various control methods are practised such as dredging the oyster grounds. One of the most effective methods is the placing of rows of curved tiles at low water. The tingles congregate under them to deposit their egg capsules. At regular intervals the tiles are turned over and the tingles and their egg capsules removed and destroyed.

4.2.2   *The American slipper limpet* Crepidula fornicata

Another pest of the oyster farmer, but for different reasons, is *Crepidula fornicata* (Fig. 4–1). These mesogastropods have adopted filter-feeding by enlarging the mantle cavity to cover almost the whole dorsal surface of the animal. An extensive, single ctenidium provides the ciliated surface both to generate the water current and to collect the food material.

When feeding, a mucous sheet covers the entrance of the mantle cavity, trapping the coarser particles. By various ciliated gutters the mucus-bound food is collected as a cord in a food groove running up to the right side of the head from where it is periodically collected by radula movements. *Crepidula* typically live attached to one another in 'strings' some 8 or 10 individuals long. As protandrous hermaphrodites the oldest and largest individuals in the string are female, the younger and smaller ones are male.

The significance of *Crepidula* as a pest of oyster beds is in its competition with the oysters for food. Very large numbers of slipper limpets can develop, up to 40 tonnes acre^{-1} on some oyster beds, quite smothering the oysters.

4.3   Vectors of parasitic disease

The digenetic trematodes, the flukes, are widespread internal parasites of vertebrate animals. The vertebrate final host is the environment for the mature adult worms which, almost without exception, include in their life cycle an asexual multiplicative phase in a molluscan intermediate (Fig. 4–3).

Practically all of these molluscan intermediate hosts are gastropods and because of the evolutionary association of man and his domestic animals with supplies of freshwater, it is the freshwater snails, of several families, which are most important to human health and economy. With our anthropocentric view of human disease it is common to refer to these molluscs as 'vectors' of parasitic disease, but truly they are passive intermediates in the life cycle of the parasite. Although the parasite may undergo several multiplicative stages in the snail (Fig. 4–3), its effects on the snail are confusing. In some, the result of infection is retarded growth and in others, it is a form of gigantism. Although there does not seem to be host mortality due to the parasite, there is a drop in fecundity which is probably due to the diversion of material to parasite nutrition (WRIGHT, 1971).

4.3.1   Bilharzia or schistosomiasis

Bilharzia is an insistent disease of human populations over wide areas of the tropical world. It is due to one of several species of trematode which infect their human host via a snail of the family Planorbidae, usually of the genus *Bulinus* or *Biomphalaria*. Eggs from the adult worm are shed into a stream or pond via the urine. Hatching out in the water to a free-swimming miracidium stage they penetrate the snail and undergo asexual reproduction in the liver (digestive gland), emerging some weeks later as cercariae. These aquatic larvae must penetrate the skin of the

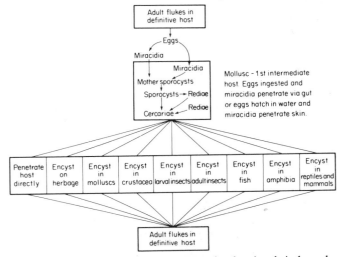

Fig. 4–3   The life cycles of the digenetic trematodes showing their dependence on the molluscan intermediate host for the multiplicative phase of their life cycle. The alternative routes to the final host are adopted by different families of trematodes. (From Wilson Olsen, O., 1962, *Animal Parasites*. Burgess, Minneapolis.)

final host within two or three days (Fig. 4–4). Moving through the blood system and lungs, the young worms lodge in the liver and grow to sexual maturity. Thereafter, adult worms and eggs can lodge in almost any tissue of the body but principally occur in the blood vessels of the bladder wall and pelvic plexuses (*Schistosoma haematobium*), or the mesenteric vessels around the gut (*S. mansoni, S. japonicum*). The eggs can penetrate the wall of the bladder or gut and are thus released in the urine or faeces into the water to renew their life cycle.

The clinical effects of schistosomiasis are varied, often difficult to diagnose and usually sub-lethal. There are transient lesions as the cercariae penetrate the skin and transitory pulmonary reactions as they pass through the lungs. As the young worms grow to maturity in the liver, symptoms such as loss of appetite, malaise, headache, fever and tenderness of the liver may occur. At sexual maturity and the release of large quantities of eggs more serious symptoms appear which are due to blockages and lesions in blood vessels of the liver, lungs and bladder. Depending on the species of worm and severity of infection these lesions may extend to the kidney, ureters and genitalia (*S. haematobium*), or intestinal walls of the caecum, colon and rectum (*S. mansoni, S. japonicum*), and may even include arterial damage in the lungs with consequent effects on the heart. Anaemia is present in all forms of bilharzia.

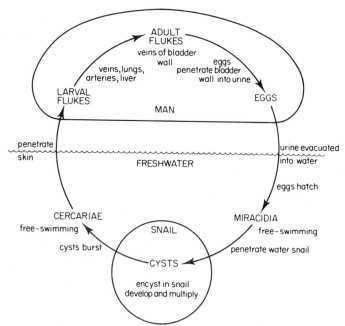

Fig. 4–4 The life cycle of *Schistosoma haematobium* (Bilharzia) in man and freshwater snails. (Adapted from *Schistosomiasis*. Bulletin No. 6, 1974, The Ross Institute, The London School of Hygiene and Tropical Medicine, London.)

The World Health Organization (W.H.O.) have major research programmes into this disease in many tropical countries. They conclude that the direct ill effects of the disease are generally under-estimated and the consequences of infection in an individual are often only apparent when a greater than normal workload is required. Thus the infected individual compensates for his condition of malaise by reduced activity. W.H.O. estimate conservatively that there are 150 million infected individuals thoroughout the tropical world and that as a cause of *morbidity*, i.e. reduction in fitness, it is probably outranked only by malaria and tuberculosis.

These human trematode parasites are virtually exclusive to man. The host restriction and their close association with planorbid snails, may be due to the parallel evolution of the parasites and molluscan intermediates with the explosive phase of primate evolution, which occurred in East Africa during the Miocene period (WRIGHT, 1971). Certainly, early man must have evolved in close association with the freshwater habitats of the snails and about 60 % of the human population of African savanna still live in areas where the snails are very common. The bulinid and biomphalarid snails favour shallow waters near the shores of lakes, ponds, streams and irrigation channels which provide suitable conditions for food, shelter and oviposition. It is exactly the same areas which are in continuous use as part of rural village life as a water supply and for washing and bathing. Children are particularly exposed and most chronic infections probably originate early in life. Agricultural improvements in the form of drainage ditches, and social customs such as the use of certain pools for ritual washing, greatly enhance the spread of schistosomiasis.

Control measures are varied in type and in success. Again, mollusci-cides based on heavy metals, arsenical or various organic compounds have been used extensively against the snail intermediates. The success of these is severely limited by their consequences upon other biological life forms and uses of the water and because the results can be only temporary. The snails are an integral part of the ecology of the freshwater habitat and measures to reduce their numbers by clearing the vegetation from which they obtain food, or canalizing the water between concrete banks to increase flow rate and reduce shelter, are not completely effective and have other drastic effects on the freshwater habitat. Breaking into the parasitic life cycle by providing uncontami-nated water for use by the human community and separating it completely from sewage pollution, the source of infection, is obviously the long-term answer. But, given the scale of the problem and the costs of this solution, it is probable that measures against the molluscan intermediates will continue to be taken for a long time yet.

4.3.2 Liver fluke in agriculture, fascioliasis

Another important area in which man is affected by snail-mediated parasitic disease is in the incidence of liver fluke in farm animals. This type of parasitic trematode occurs almost throughout the world. In Europe the adult common liver fluke, *Fasciola hepatica*, is found in the bile ducts of the liver of the mammalian host, usually sheep or cattle. The flukes are hermaphrodite and produce large numbers of eggs which reach the lumen of the host gut in the bile. They are eventually shed to the outside world with the faeces (PANTELOURIS, 1965). On moist and boggy pasture the egg hatches to a miracidium and penetrates the snail intermediate host, *Lymnaea truncatula*. After several multiplicative stages in the tissues of the snail, a free-swimming cercaria emerges and encysts on to vegetation from where it is ingested by the grazing mammals.

The effect of the infection is to cause blockages of the bile ducts resulting in 'liver rot'. The liver becomes unsuitable for human consumption, because the worm can be infective directly in humans and there is considerable loss of condition, milk and meat production from the animals. In Europe, where the disease was once a severe problem, its incidence has now been greatly reduced by rigorous abattoir inspection of livers, the use of antihelminthics in infected animals and the draining and management of pasture. The latter methods are, of course, again directed at the snail intermediate host which is essential to the transmission of the disease.

4.4 Marine fouling

Unless it is coated with preservatives or paints which are toxic to marine life, any man-made object placed in the sea rapidly becomes encrusted with marine life. This colonization of the surface and its use as living space involves a great variety of marine organisms. Eventually an epifaunal community will develop which resembles that which would be present on adjacent natural surfaces. When colonization and succession occurs on man-made constructions it is given the general term of 'fouling'.

A number of molluscs are important fouling organisms. They fall into two major categories, those which penetrate the surface material and weaken its structure and those which thickly encrust the surface. In the first category are several groups of bivalve molluscs which have evolved some ability to bore into natural and artificial structures. Two closely-related families, the Pholadidae (piddocks) and the Teredinidae (shipworms) are by far the most important commercially, especially the latter. In the second category, many bivalve families attach themselves to surfaces by cementation or byssus threads, but it is the mussels (Mytilidae) which are the most important fouling molluscs.

4.4.1 Piddocks (Pholadidae)

Boring into rock or wood is achieved in piddocks by the use of modi fied shells to abrade the material, coupled with an anteriorly directed foot to hold the shell against the rock surface (Fig. 4–5). These specialized borers show a number of adaptations to this habit, central to which is the ability to rock the shells to and fro in the long axis of the body. This necessitates the use of the hinge as a pivot with consequent loss of hinge teeth and reduction of the ligament. The rocking motion is the result of alternate contraction of anterior and posterior adductor muscles.

Piddocks bore into soft rock and mudstones and it has been known, for genera such as *Pholas, Xirphaea* and *Barnea,* to damage structures made of cement and soft concrete. Built in 1898, the La Boca Dock (Panama Canal Zone) was so extensively damaged by rock borers that in

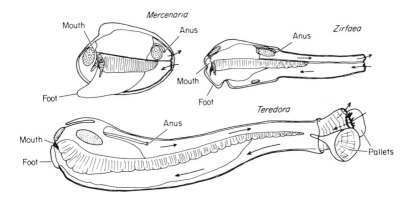

Fig. 4–5 The comparative anatomy of a relatively unspecialized bivalve (*Mercenaria*), a rock-boring pholad (*Zirfaea*) and a wood-boring teredinid (*Teredora*). (Simplified from Turner, R. D., 1966, *A survey and illustrated catalogue of the Teredinidae.* Mus. Comp. Zool. Harvard, Cambridge, Mass.)

1922 it was condemned and closed. Cement mortar piles are also attacked by pholads as well as rock borers from other families (e.g. *Platydon,* a mytilid). Some pholads may bore into wood (e.g. *Martesia*) and cause significant economic loss.

Like most bivalves the pholads remain filter-feeders. Entry into the substratum seems to serve primarily for protection. With relatively short siphons, the pholads excavate holes not much longer than the body. Damage to concrete structures in the sea is now largely avoided by the use of materials of sufficient hardness.

4.4.2 Shipworms (Teredinidae)

This family of highly specialized wood borers is of enormous historical and commercial importance. Although some filter-feeding is possible, they differ from other bivalves in using the substratum itself as food. Thus, growth, feeding rate and the amount of boring are directly linked to the scale of damage. They are one of the major groups of marine animals contributing to the reduction of wood in the marine environment and unless thoroughly protected, wooden structures of all sorts may be attacked and destroyed.

Because of their economic importance, the biology of teredinids, particularly *Teredo* in the Atlantic and *Bankia* in the northern Pacific, has been thoroughly studied. The shells are greatly reduced and each has a characteristic, nearly hemispherical shape with a deep right-angled notch in the ventral half of the anterior margin. The large, discoid, muscular foot protrudes through the gaps formed by the notches and, acting as a powerful suction disc, holds the valves against the wood at the head of the burrow. Pivoting about the hinge, the two shell valves are rhythmically rocked to and fro by alternate contractions of the anterior and posterior adductor muscles (Fig. 4–5). When the powerful posterior adductor contracts, spreading the valves anteriorly, rows of fine teeth on the outer surface of the anterior margin of the shell abrade the wood, scraping off fine particles. Ciliary tracts on the foot and mantle deliver these wood particles into the mantle cavity and mouth. The wood enters a special caecum of the gut where it remains during digestion. The cellulolytic enzymes apparently originate from the numerous and extensive glandular evaginations of the fore gut, but, as cellulose-decomposing bacteria have been isolated from the intestinal organs of *Teredo navalis* it is not clear whether it is the molluscan gut or its symbiotic bacteria which is responsible for most of the enzyme production.

Using the uncalcified larval shell, the young teredo bores into the wood surface within 24 hours of settlement, cementing the resulting wood chips into a protective covering for itself. Growth now depends on several factors such as the temperature and the species of timber bored. As it bores, the body lengthens steadily leaving the siphons open to the sea-water at the point of penetration, the visceral mass makes up about a quarter of this length and the remainder is gills and mantle (Fig. 4–5). The tunnel is lined with a chalky, calcareous secretion. The animal can become adult in 4–6 weeks at a length of about 100 mm and it is thought that the average life span may be only about 10 weeks, after which it has reached a diameter of 5 mm and a length of 100–125 mm. During this time it has functioned both as a male and as a female and destroyed a column of wood of the same dimensions as its largest size. Individual specimens have been recorded up to 4 feet (∼ 1200 mm) in length and an inch (∼ 30 mm) in diameter (*Teredo dilatata*) from piles which had been in the water for less than 2 years. A study by X-ray of the growth of *Bankia setacea* showed one individual with a monthly length increase of 122 mm, reaching 610 mm, in about five months (QUAYLE, 1959).

Damage to timbers in the sea can be very rapid and complete especially in tropical waters. Untreated piles may last only 1–5 years, but those thoroughly treated with cresote preparations under pressure may last in the region of 8–60 years. Once infested, timber is very difficult to treat effectively. In adverse environmental conditions teredinids withdraw slightly and close the entrance to their burrows with structures known as *pallets* to seal them off from the sea-water. Biochemical pathways seem to be well adapted to periods of anaerobic metabolism (LANE, 1959). As a result of this, the shipworm is very easily transported about the world, inside pieces of timber, surviving adverse conditions external to the burrow and continuing to breed wherever conditions are suitable.

A good example of the speed of borer attack is related by EDMONSON (1951) from Honolulu Harbour in the central Pacific. A dredging company laying a long sewer outfall in the harbour required a trestle to be constructed out to the reef edge for the temporary carriage of materials and equipment. Although the area was known to prone to be teredo damage it was decided that as the construction was only needed for 8 months, untreated timber piles would be used. The company lost the race with the teredinids and after only 70 days, sections of the long trestle collapsed and tipped considerable amounts of heavy equipment into the sea.

The total cost of borer attack in terms of damage and preventive measures cannot be estimated but must be enormous.

4.4.3   Mussels (Mytilidae)

To epifaunal marine species the occurrence of new surfaces in the sea on which their larvae can settle provides an increase in living space. The dense settlement of mussels and oysters is exploited to advantage for spat collection in the aquaculture industry (Chapter 3), but it is a serious nuisance when it occurs on surfaces which should remain smooth.

Fouling by mussels on fixed harbour installations is often of little importance but on ships' bottoms, fouling significantly hinders the smooth flow of water over the hull leading to increased drag, slower speeds and higher fuel costs. Traditionally, the methods for dealing with fouling of all sorts rely on coating the surface with paint-containing toxic chemicals, most frequently lead and copper. This coating must be renewed at frequent intervals by taking the ship into dry dock. Encrustation with mussels has also been a severe nuisance on the inner surfaces of pipes drawing water from the sea for cooling industrial plant such as power stations. Here, the extent of the growth is usually hidden from view and cases of severe damage have occurred when clumps of old mussels fall away and are carried further into the installation causing damage to machinery or pipe fracture. Mesh screens are used to reduce this risk and the inner surface of the pipes are periodically cleaned of fouling organisms by reversing the water flow and flushing out the system with a solution of chlorine.

A new aspect to this old problem is presented by the large fixed offshore platforms of the oil and gas industry. Many of those currently operating in the North Sea and elsewhere have become thickly encrusted with mussels down to about 10 m depth (Fig. 4–6). Because mussels are essentially a littoral and shallow water organism, those sited close to the shore are more likely to be fouled in this way.

These massive structures, some now standing in water over 200 m deep, whether of concrete or steel, are designed to have a life in the sea of upwards of 30 years. They cannot be completely protected from fouling by existing paint coatings, all of which have a relatively limited life. The fouling layer, particularly of mussels is important for several reasons. Its presence on the upper part of the submerged platform contributes

Fig. 4–6 A section of a main leg of one of the Forties Field oil production platforms in the North Sea. A strip of mussel growth has been cleared away to expose a welded joint for inspection. (Photograph by courtesy of B.P. Petroleum Development, Ltd., and Sub-Sea International.)

significantly to increased drag and thus to the forces of wave loading on the structure. This has to be accounted for in the design and must add to the construction costs, and by tightly encrusting the surface with a layer which may be 10 cm thick, that surface is no longer visible or easily accessible for inspection. The inspection procedures account for a significant part of the costs of running a platform, and can be approximately estimated overall in the North Sea alone at several tens of millions pounds annually at current prices. The cost is due mainly to the high price of offshore diving operations. With each mussel having a strong attachment by byssus threads they are not easily removed by hand and strenuous efforts have to be made to clean them from areas needing detailed inspection.

5 Basic Research Material

In this final chapter it is worth recalling the role that research on any group of animals may play in the general improvement of scientific knowledge. Frequently through chance, but also through the insight of individual scientists, a particular animal or group of animals proves especially favourable for research in a particular field of biology and this may lead directly to a rapid advance in that area. Readers will have their own opinions about which discoveries overall have been generally influential, but I have made a selection here of some of the areas of molluscan biology which can be said to be important to a wider audience than those whose main interest is the Mollusca.

5.1 Nervous conduction

Living in the surface layers of open ocean the epipelagic squids are active swimmers. They use water expelled from the mantle through the ventral funnel as a form of jet propulsion. It is an effective means of locomotion, pushing the squid along backwards in short bursts at measured speeds of up to 20 knots. Directing the water flow from the funnel sideways, relative to the long body axis, controls changes in the direction of travel and a posterior pair of fins act as hydrofoils to give stability to the movement. Squids are preyed upon by many groups of marine vertebrates such as whales and seals, several families of birds and probably a number of larger fishes. Lacking any form of protective shell, an efficient escape mechanism obviously has a high survival value.

To obtain maximum acceleration and give the best chance of escaping a predator, the squids have evolved a system of special nerve fibres to distribute nervous signals from the brain to the mantle and ensure an even muscular contraction in the shortest possible time. This is the so-called giant fibre system (Fig. 5–1) which has become one of the classical preparations of modern neurophysiology and has greatly improved our understanding of the basic mechanisms of nervous conduction and synaptic contact.

The unique features of these nerve fibres – their exceptionally large diameter and length – were accurately described as early as 1909. But it was their independent discovery in the early 1930s and the recognition of their importance to neurobiology which marked the beginning of their influence in this field. The system varies somewhat in different squids but basically it consists of three categories of interconnected fibres, the 1st, 2nd and 3rd order giant fibres. There are two 1st order giants which cross over each other in the brain and make intimate physiological contact or even fuse together at the cross-over point. The conduction of nervous signals to the mantle is thus synchronous on both sides of the

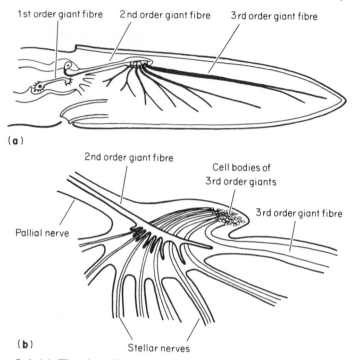

1st order giant fibre   2nd order giant fibre   3rd order giant fibre

(a)

2nd order giant fibre

Cell bodies of
3rd order giants

3rd order giant fibre

Pallial nerve

(b)

Stellar nerves

Fig. 5–1 (a) The giant fibre system of the squid *Loligo pealeii* and (b) the synapse in the stellate ganglion. (From Young, J. Z., 1939, *Phil. Trans. Roy. Soc.*, **229**, 465–503.)

animal. A limited number of 2nd order giant fibres, with cell bodies located in the magnocellular lobe of the brain, distribute the stimulation to the funnel, head retractor muscles and to the stellate ganglion, a mass of nerve cells lying on the inner surface of the mantle on each side.

In the stellate ganglion, branching terminals of one of the 2nd order fibres synapse with a series of 3rd order cells (Fig. 5–1) which distribute their axons through the stellar nerves to the complex muscle layers of the mantle. It is the graded diameters of these 3rd order fibres which finally ensures synchronous contraction of the whole mantle because the conduction velocity of each fibre is proportional to the square root of its diameter. Thus the fibre with the largest diameter runs the furthest distance in the mantle. These nerve fibres may be up to 1 mm in diameter, easily visible with the naked eye. It is the sheer size of the fibres and the synaptic arrangements in the stellate ganglion, coupled to the relative ease with which invertebrate experimental preparations may be maintained in a physiological state, which has made them so significant.

Carefully removed from the nerve and cleaned of the smaller fibres surrounding it, the giant fibre of *Loligo* provides a single nerve axon which may be several centimetres long, about 650 μm in diameter and weigh about 20 mg (Gilpin Brown in: NIXON and MESSENGER, 1977). By the late 1930s, detailed studies were being made of the electrical events which occurred on the outer surface of the nerve fibre when a nerve impulse was conducted, and the relationship between fibre diameter and speed of conduction was established. This was followed within a few years by measurements of the protein and electrolyte constituents of the axoplasm (the cytoplasm of the fibre).

A most significant step was the placing of an electrode inside the axon which was first achieved by Hodgkin and Huxley in 1939 (*Nature*, Lond., **144**, 710–11). This allowed the determination of the resting membrane potential of the fibre (due to the different distribution of electrolytes between the axoplasm and bathing medium) and also the measurement of dynamic changes, the action potentials, which are the basis of nerve impulses conducted along the fibre. An increasing range of sophisticated techniques followed and allowed the simultaneous placing of two or more electrodes within the fibre, the imposition of controlled potentials and current flow across the nerve membrane, injection of radioactive tracers into the axoplasm and even the replacement of axoplasm with solutions of altered composition. A series of experimenters discovered the relationships between the electrical properties of the nerve fibre, the ionic compositional changes within the axoplasm, and the active properties of the cell membrane in restoring the ionic balance after the passage of the action potential. A quick review of the contents pages of modern scientific journals will show that this is still a very productive field of research.

It was perhaps fortunate that the mechanism of the action potential in the squid giant fibre turned out to be a 'sodium spike'. That is, the rapid reversal of the transmembrane potential is due to the inrush of positively charged sodium ions. This is not a universal mechanism but it is certainly the case in vertebrate motor nerves so important to human physiology. The work on the squid giant axon and synapse, which because of its size has allowed the application of many physical and chemical techniques to the basic problems of nervous conduction, has thus led directly to a series of discoveries of very wide significance in biology and medicine.

5.2 The co-ordination of behaviour

Gastropods possess a nervous system in which most of the nerve cells are grouped into ganglia. These ganglia are grouped together or even fused into a circum-oesophageal ganglionic mass. In pulmonates such as *Helix* and opisthobranch genera such as *Aplysia* and *Tritonia* this concentration of nerve cells is easily accessible to experiments and has proved to be a fruitful physiological preparation for neurobiological study.

The particular feature of these ganglionic masses is that although they may contain large numbers of small cells there is a smaller number of 'giant' nerve cell bodies which are considerably larger than those of most other animals. These cell bodies can be identified individually and assigned relative positions in a cell map (Fig. 5–2). They range in size up to 1000 μm in diameter and can be easily penetrated by glass micro-electrodes to allow a continuous record of the electrical events occurring within the cell. With suitable experimental arrangements, the movements of parts of the animal such as the buccal mass, tentacles, gills, mantle or foot can be observed, at the same time as intracellular recordings are made from individually identified cells in the nervous system. Many of the fine nerves entering the ganglia can also be picked up with extracellular electrodes for recording or stimulation.

The special advantages of these nervous systems have allowed a significant variety of work concerned with understanding basic nervous function (see DORSETT, 1975; KANDEL, 1979 for reviews). The pharma-cology of neurotransmitter substances, the variety of effects due to synaptic action on the post-synaptic cell and the neural organization of certain reflexes have been intensively studied.

The recent trend of these studies has been towards investigating the neural basis of simple behavioural acts. These range from habituation of the gill withdrawal reflex in *Aplysia* to more complicated sequences of behaviour in the nudibranch sea-slug *Tritonia*. It was found that stimulation of identified cells in the ganglia caused co-ordinated motor responses. Some of these, such as a turning movement or withdrawal of the branchial tufts, simply involve a single set of effectors but the most

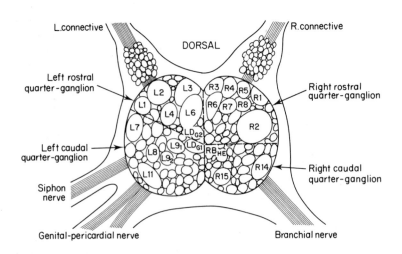

Fig. 5–2   Cell map of the abdominal ganglion of *Aplysia californica* to show some of the large number of individually identified nerve cells on the dorsal surface (simplified from KANDEL, 1979).

complex behaviour analyzed is the swimming escape response. A short electrical stimulus to one of the nerve trunks leading to the brain, elicits a stereotyped sequence of activity involving a number of different parts of the body and lasting up to 40 s. This sequence involves the reflex withdrawal of gill tufts, then further preliminary movements followed by a series of swimming movements consisting of cycles of dorsal-ventral flexions of the whole body. This behaviour, always produced in the same form, corresponds to the 'fixed action pattern' identified by ethologists in all sorts of animals. Its physiological basis in *Tritonia* resides in the linked activity, in sequence, of identified nerve cells and is thus consistent with specific neural connection models of animal behaviour.

5.3 Learning

A further area of behavioural research where molluscan examples have been influential is in the study of learning. Several levels of learning ability are recognized among animals and the simplest of these, habituation and sensitization, have been investigated in opisthobranchs such as *Aplysia* but apart from the famous exception of the cephalopods, the presence of higher forms of learning in molluscs is doubtful. The unprotected body and benthic existence of the octopuses has produced an animal in which associative learning appears to play a major part in behaviour. Broadly, this means that the octopus quickly learns to associate two or more stimuli which occur together or in close sequence such as the visual and tactile cues to the presence of crabs (food).

It was largely due to Professor J. Z. Young that the possibilities of the octopus for learning studies became widely recognized. The readiness with which the animal attacks moving objects in its tank, its acute ability to make visual and tactile discriminations and its almost insatiable appetite for food presented an ideal opportunity. The basic experiment is one in which the animal learns to associate reward (food) with a particular stimulus (such as the presentation of an artificial visual signal), and punishment (small electric shock) with an alternative stimulus. Once trained, the octopus can be tested for its memory of 'correct' responses and rewarded or punished as appropriate. Using suitable training, testing and operative techniques much has been learnt about the duration and location of memory in the octopus. Probably the wider interest in this work has been the way in which models of the memory system have been evolved and discussed.

5.4 Evolutionary mechanisms

Since the Darwin–Wallace papers of the middle of the last century evidence for the geographic variation of animal types has been an important component of ideas on the production of species by natural selection. Variation of a major taxon or a single species can be seen on a trans-continental scale or between the islands of a group or even over a few hundred metres of ground. All the major groups of animals have provided evidence of geographic variations and the occurrence of endemic species in isolated habitats. Among molluscs the land snails have been particularly studied in this respect.

There are in the order of 25 000 species of terrestrial molluscs, the majority of them pulmonate land snails. This represents perhaps a quarter of the known molluscan species and shows a successful radiation. These snails, herbivorous or carnivorous, have limited possibilities for dispersal compared to their marine relatives. Being, on the whole, visible on the surface of soil or plants and having a durable shell easily collected and preserved, large collections of land snails from around the world have accumulated in museums and private hands. Shells vary in relatively easily measured ways. Size, shape, thickness, details of surface sculpture and form of the aperture and lip, background colour and colour patterns; many of these features are quantifiable and form the basis for much of the scientific study of variation between and within molluscan populations.

Genera such as *Partula* in the mountainous islands of the Pacific provide examples of spatial isolation which, it is thought, leads to evolution of species by allopatry. Allopatric speciation is the production of distinct species from populations of animals with non-overlapping distributional ranges which thus have been genetically isolated. Although difficult to prove as the mechanism of speciation in these cases, such studies are central to much evolutionary theory.

Snail populations also show significant variation on a local scale. The classic example is provided by the studies of Cain and Sheppard (see WHITE, 1978 for review and references) on *Cepaea nemoralis* and *C. hortensis* in southern England. These snails are common in a variety of habitats and they are polymorphic (Fig. 5–3), that is, a number of classifiable colour and banding pattern varieties occur together. These variations have a genetic basis and the frequency with which various morphs appear is related to the type of habitat. Thus, the unbanded brown, unbanded or one-banded pink shells are more frequent in woodland areas but yellow five-banded forms are commoner in hedgerows. The maintenance of different frequencies of the various polymorphic forms is largely due to selective predation by thrushes.

Snail populations also show variations in the frequency of these forms within a relatively homogeneous area. These are the so-called 'area-effects' when variation in an apparently continuous population persists in the absence of obvious differences in predation, rainfall, temperature and other factors which are known to favour one or another of the morphs. The area where one variant is the most frequent may abut an area where a different one predominates without any clear geographical boundary, the change from one population to the other occurring over distances as short as 100 m. The presence of these differences within a relatively circumscribed area of land raises the important question of

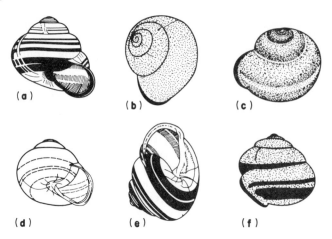

Fig. 5–3 Shell polymorphism in the land snail *Cepaea nemoralis* showing a few of the known varieties. (**a**) yellow, 5-banded, dark lip, (**b**) pink, 0-banded, dark lip, (**c**) brown, 1-banded central, dark lip, (**d**) yellow, bands and lip lack pigment, (**e**) yellow, pigmented bands, unpigmented lip, (**f**) pink, central and 2 lower bands, pigmented lip. (From Sheppard, P. M., 1958, *Natural Selection and Heredity*. Hutchinson, London.)

whether they could lead to proper speciation in the absence of physical isolation.

In at least two areas of modern evolutionary study, then, the speciation and polymorphism of land snails is providing important material.

5.5   The 'Mussel Watch Scheme'

The 'Mussel Watch Scheme' is a research project undertaken primarily to measure human activities in polluting the ocean. With these practical aims it provides a distinct contrast to the previous examples. The ability of bivalves, especially mussels and oysters to concentrate small particles and accumulate various toxins and pollutants is being used directly as a means of monitoring changes in pollutant levels. Many pollutants in sea-water may be present in significant, but extremely low concentrations, too low for many methods of direct measurement. Bivalves can concentrate these pollutants in their tissues by factors up to several million, reaching levels which can be measured accurately by the analytical chemists (BAYNE, 1978).

The scheme as it began in 1976, called for the co-ordinated and standardized measurement of pollutants in four species of bivalve, *Mytilus edulis, M. californianus, Crassostrea virginica* and *Ostrea equestris* from 107 sample sites around the coast of North America (GOLDBERG, 1978). Four categories of pollutant were measured.

Heavy metals – lead, cadmium, silver, zinc, copper and nickel, the products of a wide range of industrial processes.

Radionuclides – the radioactive elements resulting from nuclear weapons fall-out and nuclear power stations.

Halogenated hydrocarbons – this group includes the degradation products of pesticides such as DDT as well as the poly-chlorinated biphenyls (PCBs) which are widely used industrial chemicals.

Petroleum hydrocarbons – originating from oil pollution.

In addition a 'bank' or 'library' of frozen samples of mollusc tissue has been set up to allow for new advances in measurement techniques.

The scheme has immediately produced useful data on the normal levels of these substances as well as identifying pollution 'hot-spots' some of which were not previously recognized. The organizers hope that the scheme will continue and be extended to cover more sites and areas of the globe and develop into a standardized method of identifying changes in levels of these important chemicals in the sea. It is clear that whatever the particular scheme adopted the marine bivalves are likely to be of considerable future use in the monitoring of marine pollution levels.

Appendix

Phylum MOLLUSCA: unsegmented, coelomate; head, muscular foot and visceral mass; dorsal mantle, often secreting shell; paired ctenidia (gills); alimentary canal with buccal mass, radula, salivary glands; blood system well organised, heart, median ventricle with 2 lateral auricles, often extensive haemocoel; nervous system complex; coelom present as pericardium, cavities of kidneys and gonadial cavity; trochophore type of larva (veliger, glochidia).

Class MONOPLACOPHORA: bilateral symmetry, 1-piece shell, 5 pairs branched gills, 2 pairs auricles, 2 pairs gonads, nervous system primitive, e.g. *Neopilina*.

Class POLYPLACOPHORA: elongate, bilateral symmetry, nervous system primitive with longitudinal cords, 8 transverse shell plates, many ctenidia, e.g. *Lepidochiton*.

Class APLACOPHORA: worm-like with mantle investing body, calcified spicules in mantle.

Class GASTROPODA: asymmetrical, well developed head, broad muscular foot, 1-piece shell coiled helically at least in young, torsion.

Subclass PROSOBRANCHIA: usually aquatic, well pronounced torsion, nervous system visceral loop crossed in form of figure 8, mantle cavity opens anteriorly, ctenidia anterior to heart, often operculate, sexes separate.

Order ARCHAEOGASTROPODA: indications of original bilateral symmetry; primitively have 2 ctenidia, when one is present it is bipectinate, e.g. *Haliotis, Patella.*

Order MESOGASTROPODA: organs of right side of palliopericardial complex lost; ctenidium monopectinate, nervous system concentrated, free swimming veliger larva, shell may be siphonate, some carnivorous with eversible proboscis, e.g. *Viviparus, Littorina.*

Order NEOGASTROPODA: most advanced prosobranchs, very concentrated nervous system, siphonate shell, carnivorous, free swimming veliger often with suppressed intracapsular development, large bipectinate osphradium, e.g. *Buccinum, Nucella.*

Subclass OPISTHOBRANCHIA: marine, hermaphrodite, reduced shell becoming internal, signs of detorsion, uncrossing and shortening of visceral loop, return to bilateral symmetry, many orders.

Subclass PULMONATA: hermaphrodite, no ctenidium, mantle cavity vascularised as lung, signs of detorsion, shell, no operculum, nervous system symmetrical, suppression of larval stages, terrestrial.

Class SCAPHOPODA: bilateral symmetry, tubular shell open at both ends, head with prehensile proboscis, radula, no ctenidia, e.g. *Dentalium.*

Class LAMELLIBRANCHIATA (=BIVALVIA=PELECYPODA): bilateral symmetry, vestigial head, no radula, ciliary feeders using labial palps and enlarged ctenidia 2 mantle lobes enclose body, lobes secrete shell of 2 valves joined by dorsal ligament constituting a hinge, burrowing foot.

Subclass PROTOBRANCHIA: ctenidia flat, non-reflected filaments, two rows on either side of branchial axis, many primitive but some specialised characters, e.g. *Nucula, Yoldia.*

Subclass LAMELLIBRANCHIA: ctenidia large relative to labial palps to form feeding organs, filaments long and reflected, adjacent filaments may be joined by ciliary bridges (filibranch) or tissue junctions (eulamellibranch) e.g. *Mya, Mytilus.*

Subclass SEPTIBRANCHIA: adductor muscle equal, mantle edges not fused, gills change to muscular septum pumping water through the mantle cavity, mantle edges not fused, gills change to muscular septum pumping water through the mantle cavity, often abyssal, e.g. *Cuspidaria.*

Class CEPHALOPODA: bilateral symmetry, radula, 1 or 2 pairs ctenidia, coelom well developed, chambered shell, nervous system centralized, concentrated and sense organs highly organised, high metabolic rate, water circulation in mantle cavity expelled through pallial funnel for jet propulsion.

Subclass NAUTILOIDEA: all extinct except for *Nautilus*, external many chambered shell with siphuncle, head with numberous tentaculate appendages, retractile and unsuckered, funnel formed of two separate folds, 2 pairs of ctenidia and kidneys, eyes open with no cornea or lens.

Subclass AMMONOIDEA: external shell, extinct; previously very numerous.

Subclass COLEOIDEA: all modern forms, mantle naked forming sac covering viscera and a reduced shell, 1 pair ctenidia and kidneys, funnel closed tube, eyes with lens and closed or open cornea, ink sac.

Order DECAPODA: 8 arms, 2 tentacles; large coelom, suckers pedunculate with horny rims, internal shell remnant, e.g. *Architeuthis, Loligo, Sepia.*

Order OCTOPODA: 8 arms, non pedunculate suckers, reduced coelom, internal shell lacking, e.g. *Octopus, Eledone.*

Order VAMPYROMORPHA: 8 long arms united by swimming wed, 2 small tendril-like arms, fins, light organs, and pen, e.g. *Vampyroteuthis.*

Further Reading and References

AYRES, P. A. (1978). Shellfish purification in installations using ultraviolet light. *Lab. Leafl., MAFF Direct. Fish Res., Lowestoft*, **43**.

BAYNE, B. L. (ed). (1976). *Marine Mussels: their ecology and physiology*. I.B.P. no. 10. Cambridge University Press, Cambridge.

BAYNE, B. L. (1978). Mussel watching. *Nature, Lond.*, **275**, 87–8.

BOYLE, P. R. (1977). The physiology and behaviour of chitons (Mollusca: Polyplacophora). *Oceanogr. Mar. Biol. Ann. Rev.*, **15**, 461–509.

CLARKE, M. R. (1966). A review of the systematics and ecology of oceanic squids. *Adv. Mar. Biol.*, **4**, 91–300.

CLARKE, M. R. (1977) In: *The Biology of Cephalopods*. Eds M. Nixon and J. B. Messenger. *Symp. Zool. Soc. Lond.*, **38**. Academic Press, London.

COX, I. (ed). (1957). *The Scallop. Studies of a shell and its influences on humankind.* Shell Transport and Trading Co., London.

DORSETT, D. A. (1975). In: *'Simple' Nervous Systems*. Eds P. N. R. Usherwood and D. R. Newth. Edward Arnold, London.

EDMONSON, C. H. (1951). In: *Report of Marine Borer Conference*. U.S. Navy Dept. Civil Engineering, Research and Evaluation Lab., California.

FAO (1977). *Yearbook of Fishery Statistics. Catches and Landings*, Vol. **44**.

FRETTER, V. and GRAHAM, A. (1962). *British Prosobranch Molluscs*. Royal Society.

GOLDBERG, E. D. *et al.* (1978). The mussel watch. *Envir. Conserv.*, **5**, 101–26.

GULLAND, J. A. (1971). *The Fish Resources of the Ocean*. Fishing News (Books) Ltd.

HALSTEAD, B. W. (1978). *Poisonous and Venomous Marine Animals of the World*. Revised edition. Darwin Press.

HANCOCK, D. A. (1962). Spotlight on the American Whelk Tingle. *Lab. Leafl., MAFF Direct. Fish Res., Lowestoft*, **2**.

HUNTER, P. J. (1978). In: *Pulmonates*. Vol. **2A**. Eds V. Fretter and J. Peake. Academic Press, London.

JACKSON, J. W. (1917). *Shells as evidence of the Migrations of Early Culture*. University of Manchester.

JONES, E. B. G. and ELTRINGHAM, S. K. (eds) (1971). *Marine Borers, Fungi and Fouling Organisms of Wood*. O.E.C.D.

KANDEL, E. R. (1979). *Behavioural Biology of* Aplysia. Freeman, San Fransisco.

KORRINGA, P. (1976). *Farming Marine Organisms Low in the Food Chain*. Elsevier, Holland.

KORRINGA, P. (1976). *Farming the Cupped Oysters of the Genus* Crassostrea. Elsevier, Holland.

KORRINGA, P. (1976). *Farming the Flat Oysters of the Genus* Ostrea. Elsevier, Holland.

LANE, C. E. (1959). In: *Marine Boring and Fouling Organisms*. Ed. D. C. Ray. Friday Harbour Symposium, University of Washington Press.

LOVELL, M. S. (1867). *The Edible Molluscs of Great Britain and Ireland with Recipies for Cooking Them*. London.

MEAD A. R. (1961). *The Giant African Snail: a problem in economic malacology.* University of Chicago Press.

MEAD, A. R. (1979). *Economic Malacology with particular reference to* Achatina fulica. In: *Pulmonates,* vol 2B. Eds V. Fretter and J. Peake. Academic Press, London.

MORTON, J. E. (1967). *Molluscs.* 4th Edition. Hutchinson, London.

NIXON, M. and MESSENGER, J. B. (eds) (1977). *The Biology of Cephalopods.* Symp.

PACKARD, A. (1972). Cephalopods and fish: the limits of convergence. *Biol. Rev.,* **47,** 241–307.

PANTELOURIS, E. M. (1965). *The Common Liver Fluke,* Fasciola hepatica L. Pergamon, Oxford.

PURCHON, R. D. (1977). *The Biology of the Mollusca.* 2nd Edition. Pergamon, Oxford.

QUAYLE, D. B. (1959). In: *Marine Boring and Fouling Organisms.* Ed. D. L. Ray. Symp. Friday Harbour, Seattle. University of Washington Press.

REAY, P. J. (1979). *Aquaculture.* Studies in Biology no. 106, Edward Arnold, London.

RUNHAM, N. W. and HUNTER, P. J. (1970). *Terrestrial Slugs.* Hutchinson, London.

SALVINI-PLAWEN, L. V. (1980). A reconsideration of systematics in the Mollusca (Phylogeny and higher classification). *Malacologia,* **19,** 249–78.

SOLEM, G. A. (1974). *The Shell makers.* Wiley – Interscience.

WHITE, M. J. D. (1978). *Modes of Speciation.* Freeman, San Fransisco.

WILBUR, K. M. and YONGE, C. M. (eds) (1964 & 1966). *Physiology of Mollusca.* Vols 1 & 2. Academic Press, London.

WRIGHT, C. A. (1971). *Flukes and Snails.* Allen and Unwin, London.

YOCHELSON, E. L. (1978). An alternative approach to the interpretation of the phylogeny of ancient mollusks. *Malacologia,* **17,** 165–91.

YONGE, C. M. (1966). *Oysters.* 2nd Edition. Collins, London.

YONGE, C. M. and THOMPSON, T. E. (1976). *Living Marine Molluscs.* Collins, London.

Subject Index